Essays on Life, Science and Society

Shaw M. Akula

Essays on Life, Science and Society

The World through the Eyes of a Life Scientist

Shaw M. Akula
Microbiology & Immunology
Brody School of Medicine at East Carolina University
Greenville, NC, USA

ISBN 978-3-030-28774-0   ISBN 978-3-030-28775-7   (eBook)
https://doi.org/10.1007/978-3-030-28775-7

This Springer imprint is published by the registered company Springer Nature Switzerland AG
The registered company address is: Gewerbestrasse 11, 6330 Cham, Switzerland

Dedicated to my beloved wife, Chitra Venkatesan.

Foreword

Shaw asked me some time ago to write the foreword for his next book. I've known Shaw for over 20 years. The two of us trained together in graduate school, have collaborated scientifically on multiple projects and scientific manuscripts, and have been good friends for a long time. I naturally agreed. At the time, I had assumed it would be some form of scientific text. A few years later when he sent me the rough draft of *Musings of a Scientist* (the original title for this book), it took me a little by surprise. My first thought was something along the lines of "It seems to be a combination of Science and Philosophy. Scientific Philosophy? Interesting! Do we scientists do Philosophy?" Science and Philosophy are two things that you so seldomly think of putting together. When you think of Philosophy, it conjures up images of an old man, wrinkled from age and wisdom, sitting at the pinnacle of some mountain in some obscure area of the Himalayas, where you take the long arduous journey to ask questions such as "What's the meaning of life, the universe, and everything?" or someone who spends years of contemplation to come to the conclusion, "I think, therefore I am." Throw a scientist into that mix, and you have a completely different recipe. Your average scientist would have spent 15 min of contemplation and come up with the conclusion, "Think or else you're nobody."

When you invite a scientist to a philosophical debate, you had better come prepared with the facts. To debate with a scientist using esoteric, abstract ideas is metaphorically like bringing a knife to a gun fight. The scientist will tend to drag you back down to Earth and beat you to death with the facts. This is because that's what scientists do for a living. When you think of a scientist, you think of the person that has all the answers. It conjures up images of a person, probably with wild unkempt hair, who has a mind like an encyclopedia, spewing out information on a broad spectrum of topics, and the person in all the action and science fiction movies whom the hero or heroine goes to in order to find out how to save the world. This is in fact only a distorted version of the truth. Scientists do acquire a vast amount of knowledge because it's a requirement of the job. Science is a very competitive arena and knowledge is power; "Think or else you're nobody." A majority of scientists are motivated by the desire to make the world a better place in one way or another; they'll just spend most of their lives doing it rather than doing it dramatically at the end of an hour and 50-min blockbuster. We're also frequently inundated with people coming to us for answers and insight.

A part of being a scientist is being a teacher, both inside and outside of work. When people find out you're a scientist, you get asked for insight and answers to a wide variety of questions: "Should I get the flu shot this year?" "What is intelligence?" "Can you explain GMO products to me?" "Is there any basis to global warming?" "What came first, the chicken or the egg?" "What's true, Creation or Evolution?" (Now that I think more about it, maybe as scientists, we often do venture into the realms of Philosophy.) My personal favorite was the time I was asked whether it was true that a major chicken franchise no longer used real chicken but rather used a genetically modified chicken that had only breasts and legs. In general, scientists tend to be very analytical, to have slightly larger than normal egos, and to have a sense of humor (if you don't believe that scientists have a sense of humor, try reading Einstein's *Relativity* or Stephen Hawking's *The Universe in a Nutshell*). My answer was that yes, what most people thought of as just a chicken franchise was indeed one of the most advanced, cutting-edge bioengineering companies in the world and that indeed they had generously chosen to provide the world with low-cost chicken rather than giving in to corporate greed and using their knowledge for the more lucrative production of organs for transplantation. I received the shocked reply, "Wow! Is that really true?". To which I replied, "No, it's really just chicken. But that would be so cool if it was true." He thought about it a minute and said, "I guess that would be kind of cool to be able to produce real food without the plants or animals. Kind of science fiction like."

Some of my colleagues and I were talking one day about these continuous requests for advice and knowledge and swapping stories about the most bizarre questions we had ever been asked. One of our conclusions was that so very often, the person asking the question had never taken the time to think about what they were asking. They were not in the habit of questioning what they believed and why they believed it. I believe this was the intention of Shaw's book, *Essays on Life, Science, and Society*. I don't believe Shaw intends for everyone who reads his book to take what he says simply at face value and agree. Scientists are very much about debate, the honing, and refinement of what you believe in the context of the facts. Our beliefs and opinions as scientists continuously change and evolve in the face of new evidence. While I agree with many of Shaw's conclusions, I must admit that I don't necessarily agree with all of Shaw's premises. This of course is not surprising. There's a kind of an inside joke among scientists that if you give three scientists one data set, you'll get back four different conclusions. I don't believe that agreement is the point of this book, nor is it the point to decide whether you agree or disagree and then move on. The challenge is to apply reasoning and facts to what you believe. I agree or disagree with Shaw's conclusions, because .... This book provides you with a framework of reasoning behind a wide range of topics and challenges you to seek out the facts and details for yourself and truly understand why you believe what you believe and whether you agree with Shaw's conclusions or not. I thoroughly enjoyed reading this book. I hope that it stimulates you to think about things that you had never thought of, delve into topics that you didn't know existed, and gain a better understanding of what you yourself actually believe.

Sincerely

Philadelphia, PA   Adrian J. Reber

Disclaimer

This book is intended to benefit inquisitive minds across all the boundaries and not to hurt anyone's sentiments.

Acknowledgment

My sincere thanks go equally to Dr. Lia R. Walker and my sister, Ms. Anu Russell, for their valuable edits.

Contents

1

Understanding Order in the Disorder

Which came first? The chicken or the egg? This paradox has been argued in multiple forms since time unknown. I have wondered about it myself before concluding in favor of the egg. One may argue that this is based on the neo-Darwinist approach that states DNA information contained in an egg is what is crucial in giving life to a chicken (or for that matter any life) and that chicken is a mere mortal carrier of the genetic code. This may well be proven wrong in the near future or debated by a theologian but the fact remains that the single cell zygote in the egg generates to form the chick. So robust is the building block of our existence that it has kept life going for 3.5 billion years and more.

Life is captivating in all its exclusive splendor. At times, it can be confusing as to what life depends on, who defines it, and how. For a biologist, the smallest building block of life is a cell. While, for an astrophysicist or biochemist, elements like carbon, hydrogen, oxygen, nitrogen, and others make up life. Interestingly, both are correct in their own capacity. In reality it is cyclical, physics informs chemistry and chemistry informs biology to take us closer to understanding the phenomenon of life.

I can only ideate the thrill a young student experiences while observing cells scraped out of their own buccal cavity. I went through the process in 10th grade, and at that moment, I knew I was observing a piece of me under a light microscope for the first time. It was a mixture of excitement and curiosity, for I was looking at the smallest building block that made me. The structure, size, and function of a cell may vary depending upon histology, but the fact remains that these cells make up what we are in real life. Cells define tissues and tissues integrate appropriately to form organs that in turn

© Springer Nature Switzerland AG 2019
S. M. Akula, *Essays on Life, Science and Society*,
https://doi.org/10.1007/978-3-030-28775-7_1

collaborate in a fitting manner to give life to the form we observe around us. Order is hard to see even among highly evolved primates. Who's to be trusted? You never know when one is going to invade, plunder, kill, steal, rape, or provide a helping hand. It is chaotic beyond imagination, and sadly, that is the state in which we live. In spite of this, there is actually a tremendous level of order in our building block, the cell.

As a molecular biologist, I use tools to generate proteins, *in vitro*, on laboratory benches. Albeit, these molecular biology tools are far less efficient compared to the cells; efficiency in here is discussed in terms of the quantity produced and reproducibility. Many times, I have been baffled by the order and organization in cells. Why do we need such a tight regulation at the cellular level? Isn't it a perfect paradox when such orderly cells give rise to completely disorganized higher forms of life? Can we compare cells that make up plants and animals to sand or something finer? Are these organismal building blocks analogous to the smallest of building blocks used to build houses? What happens inside a cell? To date, I have been asked a zillion questions about cells and what they do, but, presumably, the majority of the population goes on with their lives without thinking about the small dynamic blocks that make up this mesmerizing world.

Herein, I would like to define what cells are in simple terms while also trying to give an idea about what needs to be done to understand these wonderful micro-living blocks. Understanding the cell in its entirety will bring us one step closer to more effectively treating a variety of disease conditions—a possible prerequisite for lifespan extension. Philosophically, I believe that understanding such orderly cellular behavior could quite frankly serve as a 'how-to' guide for reducing the disorderly behavior amongst humans.

When one talks about a cell, our system of general education informs us that it is only the business of a biologist to involve in educating themselves about the cell, and I believe that it is not the correct approach. There are as many facets to a cell as there is to life. First, I am going to clarify a cell in terms of biology and chemistry. There is order to the highest levels noticed microscopically within cells. This is critical as they define a specific tissue. Tissue is an ensemble of similar cells derived from the same origin meant to perform one specific function. For example, hepatocytes are cells that make up most of the tissue that forms the liver. These hepatocytes regulate key functions like detoxifying various metabolites and protein synthesis. It is because of this that the liver is one of the 4 important specimens collected by a forensic pathologist, as this organ manifests poisoning during an autopsy. Basically, specific cells make up distinct tissues that form unique organs like kidneys, brain, blood vessels, and so forth. This is one of the main reasons for why an order among cells is imperative. If this order ceases to exist, life is impossible in the form and manner we observe it today.

What are Cells and what do they do? That is the most common questions asked to me by many friends and young children, including my daughter. My simple answer is a cell is like a home (not a house). Though it is the building block of all life forms, it would be a sacrilege to compare a cell to a brick, a common building block of a house. This is primarily because bricks only provide the structural or aesthetic component of a house while cells go multiple steps ahead by defining a house into a home. Let me explain this in the following passage. A dynamic and unique layer of membrane referred to as the cytoplasmic membrane serves as the outer layer of the cell. This can be compared to the doors and windows in a home. The cytoplasmic membrane has a key role in selectively letting chemicals, elements, enzymes, or even pathogens enter cells. Along the same lines, it also allows elements, enzymes, proteins, and even pathogens to pass out of the cell. By this way, the cytoplasmic membrane behaves as a door. It also allows for exchange of gases and by this way, it functions as a window too. Once inside the door, you arrive in the living room, which would be equivalent to the cytoplasm, the amorphous fluid layer. This is where any agent that has entered the cell gets to relax a bit before being sorted to different destinations based on the information they carry. For example, a steroid molecule would be shunted off directly and immediately to the nucleus, whereas one of the dreaded viruses such as a polio virus, will be retained in the cytoplasm itself (wherein it may start to replicate).

As I said before, a cell has multiple compartments like a home: a study, kitchen, generator room, bathroom(s), storage, and a security system as well. The nucleus of the cell which is anatomically placed in the center of a cell is a membrane bound organelle. This is the study room that contains not only the critical genetic information but is also the place for the initiation of all signals that trigger the function of a cell. Right outside the perforated nuclear membrane is the endoplasmic reticulum and ribosomes that generate new proteins. These newly generated proteins are processed and packaged in a manner to be functionally active by the golgi apparatus. The endoplasmic reticulum, ribosomes, and the golgi form the kitchen.

The generator room is the mitochondria that supplies the cell with the energy needed to sustain itself. As a modern home, there are multiple bathrooms which are analogous to cellular lysosomes and peroxisomes. Lysosomes contain digestive enzymes that breakdown worn-out organelles, food particles, extraneous agents such as viruses and bacteria. Peroxisomes help the cell get rid of toxic materials. A cell also utilizes a very complex system referred to as the ubiquitin-proteasome complex to recycle proteins. All proteins are composed of amino acids that can be recycled but not synthesized by our body. The ubiquitin-tagged proteins are systematically broken to the smallest units and recycled back to synthesize new proteins. It works like the recycling machinery of the universe, the black hole, as described by Stephen Hawking. As a backup, the cell stores extra food, supplements, and water in vacuoles that can be tapped into during times of necessity. Within a cell, microtubules (a part of the cytoskeleton system) play the part of a good host, assisting in the movement of agents, proteins, and particles within a cell. For example, her-pesviruses utilize a network of microtubules to specifically traverse from the cytoplasm to the nucleus. Such an elaborate block of cell also has a robust security system which is often times broadly referred to as the major histo-compatibility complex (MHC). Additionally, other receptor molecules behave as door-keepers, standing right outside the door (the cytoplasmic membrane). They protect the sanctity of a cell. Each type of cell has a unique structure. For example, the structure of neuronal cells is distinct compared to cells that make up striated muscle fiber. This structural uniqueness is provided by cytoskeletal elements (microtubules, microfilaments, and actin) that also aid in any form of cellular movements and transport of materials within a cell. I was able to find a homologue for every room in a home to a compartment within a cell, except the dining area. Apparently, a cell is selfless and all it does is to feed and sustain life on a platter; isn't that a great character?

An appreciation for the biology of cells till this point acts as a segue into the world of medicine. The extent of the practice of medicine is an absolute reflection of what we currently know of cells. However, from hereon there seems to be a disjoint in the progress made in understanding the biology of cells and the manner it is being adapted to define and-re-define the art of clinical medicine. It is all about improving medicine for the future practice of medicine. So, the obvious question is: 'How do we go about improving medicine for the future?' The gold standard of clinical practice followed in this era is often times referred to as the "evidence based medicine or practice (EBP)". This is based on three fundamental principles: (i) research evidence on how and why a treatment works; (ii) clinical expertise on how the patient responds to a particular treatment; and (iii) patient preferences and values. In the rest of my discussions, I will focus primarily on the foremost principles of how research to understand cell biology will benefit both medicine and mankind as a whole.

Why is this crucial? Aren't we doing fine with modern day medicine? The answer is a big: "NO!" I would say, we have plateaued with our advancements in medicine for quite some time now. Our understanding of biology must feed developments in medicine. Sadly, that is not what is happening now. Modern day medical advancements are purely minor modifications being constantly done on already available physics-based equipment, such as the use of different scanners. Medicine has definitely made the retrograde transition from being a service oriented healthcare system to a money fleecing multimillion-dollar industry. However, modern medicine is still archaic in many ways. To keep it short, I am just going to provide two examples out of the scores available: (i) a mid-wife can still successfully deliver a baby as efficiently as a gynecologist unless there is a necessity for a surgical intervention. The only difference being that the midwife can deliver the baby for a pittance compared to a gynecologist. What has medicine done to diagnose a possible "cord accident". Umbilical cord accident is a major cause for 13–14% of stillbirths around the world; (ii) we still do not have a cure for simple pollenbased allergies that occur year after year. We cannot place the blame on doctors for any of this, as they only prescribe medicines; it is the job of scientists to develop medicines. However, scientists are restrained due to lack of both resources and knowledge, such as limited knowledge on the biology of immune cells with relation to allergies. To really make effective medication for the future, a good understanding of biology and other branches of science is a must.

At the outset, my description of a cell may paint it as a happy and easy going home. However, at a microscopic level it is complicated beyond imagination. Let me give you an example by describing the complex protein recycling system within a given cell. The process of recycling proteins within a cell involves three major enzymes referred to as E1, E2, and E3. In humans, there are about two E1 enzymes, 50–60 types of E2 enzymes, and between 600 and 800 E3 enzymes giving it potentially 90,000 different enzyme combinations necessary to perform cellular functions. Interestingly enough, this system is being targeted to develop effective anti-cancer drugs. The environment within a cell is highly dynamic and at the same time neatly partitioned. At any point of time, there are multiple events occurring within one given cell like,

(i) Several genes are being expressed to make proteins (there are about 20,000 genes within a human cell),
(ii) Expression of several genes are being down regulated,
(iii) Enzymes are being translocated between compartments within a cell,
(iv) Proteins being made within a cell are excreted outside to affect other cells,
(v) Energy is being synthesized to power cellular machinery,
(vi) Broken down products within a cell are being recycled,
(vii) The cell is being confronted by extrinsic agents like bacteria, fungi, viruses, and so forth.

These are only examples of a few of the different activities taking place within a cell. The actual mechanism is far more intricate. For example, if you hear someone talk about transcription of a gene, it basically means the crucial and first step of expressing a gene to make a specific protein. This step involves a segment of DNA (a particular gene) being copied to RNA by an enzyme RNA polymerase. However, this is not all. Other factors must be considered during the process of successful gene transcription:

(i) What signal determines that the gene should be copied to RNA?
(ii) There are several types of polymerases, what signal determines the choice of polymerase to actively copy DNA to RNA?
(iii) What signals the polymerase to only copy this gene?
(iv) What feedback signal prevents a secession of that particular process?

How these functions are regulated is via a bunch of proteins referred to as cellular signaling molecules (protein kinases or phosphatases), apart from other chaperones. In the last 20 years or so, we have started to understand the roles of these molecules in regulating functions of cells. However, our appreciation for this mode of regulation is far from being complete.

Our understanding of cells has been slowly but progressively changing over the years since it was first described by Robert Hooke in 1665. Our medical practices can get superior with a good grasp on what is happening within a cell. Treatment modality is only dependent on our understanding of cellular mechanisms. Over the years, we have bettered screening protocols, diagnostics, treatment, and surgery procedures, and this has had a positive correlation with new revelations about the cell and life itself. Let me give you a perfect example of how treating cancer has evolved over the years. Until the late 1980s most of the cancer treating regiments were aimed at killing cells that were actively dividing faster than the normal, healthy cells. This approach reflected our understanding of cancer at that time. With time, scientists unearthed the fact that this increase in unregulated cell division noticed in cancer cells was driven by the milieu provided by certain specific cellular signals. Knowledge of cell signaling allowed scientists to develop targeted therapies that were meant to destroy cancer cells. Targeted therapies work by influencing the actual processes or the mechanisms that control growth, division, and the spread of cancer cells, including the cellular signals that cause cancer cells to die naturally. Targeted therapies include the use of growth signal inhibitors, angiogenesis inhibitors, and apoptosis-inducing drugs. Thus, in the last three decades cancer treatment has changed from directly killing those fast replicating cells to actually blocking the cellular mechanisms that trigger such an unregulated growth. But, this is not yet complete as scientists are far from finding the magic bullet to cure cancers. As an optimist, scientists take time to unravel the mysteries surrounding the functioning of a cell. Considering the limited availability of knowledge on the topic, scientists have made quite good progress in the last three decades.

Therefore, to develop next generation screening, diagnosis, and treatment approaches, it is imperative for us to unify our understanding of what is life; more from a point of the cell. We need a unified theory to explain life in terms of a cell, the building block of life. I have always wondered how to unify biological themes. After a lot of thought, I came up with these three systematic steps to unify the cell theory and they are:

(i) To understand the effect of a particular act within the cell. For example, what happens when the cell is deprived of its nutrition? It dies; to an extent, such basic modalities have been well documented.

(ii) To understand the signaling molecules that transmit signals within a cell to regulate each of the effects observed in a cell. For example, we must have an appreciation for cell signaling that triggers self-destruction of a cell without hurting the neighboring cells when invaded by a pathogen. Scientifically, this is referred to as 'apoptosis' or the pre-1990s term, 'cell death.'

(iii) It is imperative to understand the building blocks of a cell.

For the most part, science has delineated the direct effects observed within a cell. However, points (ii) and (iii) are far from being completely unearthed. The first description of a kinase (trypsin-kinase) was published in 1931 by J. Pace. However, it took exactly 60 more years to get an appreciation of the well-known cell signaling cascade referred to as MAPK. The first published work on MAPK signaling (one of the multitudes of pathways) was in 1991. Honestly, it was a nightmare to me when I came to the US from India to pursue my graduate education in 1994; primarily because I had no clue about cell signaling. It took me a while to get a grasp on what it was and how it affects each and every function of a cell. Understanding cellular signaling is still a work in progress as we discover newer facts about signaling every day. The branch of cell signaling has unearthed quite a few biomarkers and targets that are being used to diagnose and treat a variety of disease conditions effectively. Basically, these biomarkers allow a clinician to predict a pathological condition at a relatively earlier time point. This is because any cellular manifestation is a summation of the activities of the cell signaling molecules. However, there are two major issues when it comes to adapting basic science to clinical medicine:

(i) Deciphering the function(s) of a gene is a different ball game compared to elucidating the role(s) of signaling molecules. One gene encodes one protein; however, one protein may regulate one or more cell functions. Furthermore, if the protein happens to be a signaling molecule, it is going to be associated with several cascades regulating multiple cell functions. Inherently, there is quite a bit of redundancy between signaling pathways mediated by one protein. Let me give you an example: protein p38 can regulate cell survival, apoptosis, migration, differentiation, metabolism, and proliferation. This p38 signaling protein can be activated by a variety of external stimuli like growth factor(s), inflammatory cytokine(s), oxidative stress, UV light, and DNA damage via specific and different secondary messengers like kinases, GTPases, and adaptors. To make things more complicated, p38-induced signaling significantly alters the effects of Akt, NFkB, Erk1/2, JNK, and other-associated signaling pathways. Mind you, this is only one protein and there are hundreds of other kinases and phosphatases involved in host cell signaling.

(ii) Lack of effort to create a library of signaling pathways which would provide a holistic effect within a given cell, tissue, or a condition.

Ideally, the best way to combat these issues is for federal funding agencies to take it upon themselves to create a roadmap of signaling in terms of physiology versus pathology. If this is not effectively established, the benefits of understanding host cell signaling will not be valuable enough to design drugs or for that matter the next generation diagnostic tools.

So, why is understanding the building blocks of a cell of such great importance? For a biologist, a cell is the building block of life. Also, a biologist frequently and proudly ends up describing how well coordinated things work in a cell. Order is the word to describe activity within a cell. I have been harping on the infinite degree of order found in the cells for quite some time now. For a biologist, this seems to be a perfect example of how physics laws of entropy work. The nineteenth century physicist from Austria, Ludwig Boltzmann, derived the famous equation, $S = K. \log W$, that describes the universal concept of entropy. It states that there is a trend for everything in the universe to move from order to randomness. Ah, maybe this would explain how the orderly cells give life to its seemingly endless variety or disarray.

But, remember, a cell is an entity that is further divisible into smaller units as described by physicists. Isn't it a biologist's responsibility to understand the dynamics of the smaller units that make up a cell? What are these smaller units that function as the building blocks of a cell? What sort of an order exists at that sub-atomic level? How do we understand these complex interactions? How will this benefit the future of science and the practice of medicine? To understand all of these, a biologist has no choice but to depend on the laws of physical principles. In 1944, Erwin Schrodinger, a physicist, argued that the hereditary molecule (DNA) must contain a coded-script critical to an individual's development. Believe it or not, both Watson and Crick have admitted separately how Schrodinger's idea inspired them to solve the structure of DNA which occurred in 1953. This is a perfect example of how physics helps biologists.

In layman's terms protons, neutrons, and electrons can be used to build an atom or even an entire universe. Different combinations of these sub-atomic particles give rise to diverse elements like gold, silver, copper, and other things we see around us including wood, sponge, rubber and so forth. By the twentieth century, this list of sub-atomic particles grew enormously as we started identifying new particles from the cosmic rays. Over the years, these particles were grouped into distinct families based on their mass, life, spin, and charge. The birth of the standard model of particle physics was when three new fundamental subatomic particles were classified as,

(i) The fermions comprised of the six quarks (up, down, charm, strange, top, and bottom) and six leptons (electron, electron neutrino, muon, muon neutrino, tau, and tau neutrino).
(ii) The guage bosons fundamental force carriers (weak: W and Z; electromagnetic: photons; strong force: gluons). The gluons cement the quarks in the neutrons and protons and thus hold the nucleus together.
(iii) The Higgs boson particle also melodramatically referred to as The God particle was confirmed to exist in 2013 using a high-energy particle accelerator at Geneva.

In actuality, the combination of guage bosons construct atoms using fermions in a medium (Higgs boson medium) made up of the Higgs particle to give rise to distinct elements that go on to shape the entire world as we see it. At this sub-atomic level, it is a state of chaos compared to the order observed within a cell. There is no chance of exactly identifying an event with respect to the sub-atomic (quantum) particles. At best, we can only predict the probability of the occurrence of an event with respect to these particles. All the more, at that level of size, quantum entanglement, a phenomenon that only a magician can perform on a stage in real life, makes it even more complex for a biologist to appreciate acts of quantum particles. This is where a biologist will face a tough resistance to the prevailing dogma of the cell and molecular biology's understanding of life. All that he/she knows of the well celebrated four DNA bases that weave a maze of the entire realm of the spooky complex and entangled genome that give life a form will pale into insignificance and seem simple before the perplexity of the quantum particles. It is a beautiful paradigm where the most disorderly behaving quantum particles produce an orderly cell which in turn fuels a superlative complex phenomenon of utter randomness called life. The universal law of entropy states that orderly states move toward disorder. But laws of physics do not change with time. A disordered state can also give birth to an orderly one. But that is a rarity in life. This may well be the perfect example to a physicist where from a virtually high entropy state (subatomic particles) nature provides a state of low entropy (cell) which in turn creates a state of increased entropy (life). To my knowledge, this is an example to demonstrate that the law of entropy may work in both directions.

To understand the manner in which atoms and the sub-atomic particles build a cell is critical, but as biologists, we need help from our fellow physicists. But why is this study relevant? Such a study of how atoms build a cell is critical to our understanding of the difference between life and death in terms of sub-atomic particles. What is the contribution of sub-atomic particles to the sustenance of life? What happens to subatomic particles upon death of a cell? Do these sub-atomic particles influence the whole concept of life? Such a study, quantum biology, could well be the next necessary leap in biology as we transition through the twenty-first century. I earnestly believe that a unified biology describing life in terms of its smallest building block will open new avenues in treating pathological conditions effectively and possibly improving our life on earth. The biology of the future must include systematic collaboration between traditional biologists, signal transduction experts, and quantum biologists. So, it is time for the zoologists, botanists, cell biologists, microbiologists, immunologists, and clinicians to accommodate quantum biologists into their fraternity in an effort to unify biology themes for the betterment of mankind as we transcend into the years to come.

2

Einstein's Theory of Relativity Borrowed from the Spice Rack of Hinduism?

I was born from humble beginnings and raised in a small township located on the outskirts of a City with Hindu religious importance in South India, Tiruchirappalli. Here, I grew up in a predominantly Hindu society along with its indigenous cultural values. Though I was never forced to learn Hindu scriptures at home, I grew up surrounded by its' heavy influence all around me. In spite of not being very religious, I was reminded incessantly of certain Hindu teachings and beliefs by my teachers, parents, grandparents, relatives and friends on a daily basis. At their core, they convey the same meaning as some of the common phrases used in day-to-day life:

(i)   What has to happen will happen!
(ii) This is not your fault, just fate!
(iii) If you do good, good things will happen!

These above sentences are examples of a few sentences that every Indian growing up in India would have experienced regardless of their religious affiliations. If you observe closely, the third phrase is a direct contradiction to the first two phrases. How can good things happen to someone because of their good deeds, if they indeed have no control over their lives? Oh Well! This never-ending chicken-egg argument is for another day. In any case, I have always maintained that these timeless quotes were nothing more than different excuses passed on from one generation to the next and thrown into conversations as situational decorum. The curiosity still piqued me from time to time and I wanted to trace the origin of these commonly used phrases: The

© Springer Nature Switzerland AG 2019
S. M. Akula, *Essays on Life, Science and Society*,
https://doi.org/10.1007/978-3-030-28775-7_2

first phrase is a colloquial version of an excerpt from Bhagwat Gita; one of the holy Hindu scriptures. The original version reads as follows:

> "Whatever happened in the past, it happened for the good; whatever is happening, is happening for the good; whatever shall happen in the future, shall happen for the good only. Do not weep for the past, do not worry for the future, concentrate on your present life." The basic meaning of these lines is that a supreme power is controlling the universe on a macro-scale and is also controlling the lives of every person on this earth on a micro-scale.

The second phrase is derived from the preaching of Lord Matsya who explains that three elements—fate, effort, and time—conjointly affect the course of one's life. The variable 'fate' is a powerful one where "man" has no control over the cards being dealt, so to speak. 'Fate' may refer to the hand of God.

The third phrase regarding good things happening in response to good deeds is basis of the concept of karma. Karma is the central dogma of three old-age religions: Hinduism, Buddhism, and Jainism. However, I must admit the seed for this thought was sown in the Upanishad of the Hindu scriptures [seventh century BCE]. The following lines from Upanishad preaches karma:

> *Now as a man is like this or like that,*
> *according as he acts and according as he behaves, so will he be;*
> *a man of good acts will become good, a man of bad acts, bad;*
> *he becomes pure by pure deeds, bad by bad deeds;*
>
> *And here they say that a person consists of desires,*
> *and as is his desire, so is his will;*
> *and as is his will, so is his deed;*
> *and whatever deed he does, that he will reap.*

Interestingly, all these sermons were preached a few hundred years before the modern-age religions even came into existence. For some reason, I strongly believed in the thought that good things happened in response to good deeds. On the contrary, my faculty of reasoning strongly revolted against the ideology that "what has to happen will happen". Partly because I strongly believe in the fact that I, not fate, dictate my reality.

Growing up, I had multiple subconscious debates and "pooh-poohed" on the theological idea of "what has to happen will happen," and moved on with life. It was around the time I turned 23 that my father introduced me to "Theory of relativity" by Sir Bertrand Russell. The book was a simplified version of the original description by Einstein. My dad warned me that if I were seeking understanding, a one-time casual read would do the contents of this book no justice. True to his suggestion, I had to read the book at least 5 times before my mind could amalgamate a portion of Einstein's theory. To me, having understood the general theory of relativity, was a great achievement. This was followed up by multiple brush ups on my concept of Einstein's theory of relativity, as I valued it more than anything else I understood in my life.

About 25 years later, I was on a long drive heading home after a trip and it was close to 11 at night. As opposed to the "normal" practice of focusing on the road ahead, my mind was drawn to the night sky littered with stars. I began thinking about how different astrology (as practiced by the Indians) was as compared to astronomy. Indian astrology dates back to the seventh century. Whether one believes it or not, it is mind boggling to have developed a field of study on those lines. Without any relevance, I reasoned out a perfect coincidence between the great Indian belief and astrophysics; specifically, the theory of relativity. To understand the excitement, I had to explain this epiphany to myself from the standpoint of space and time as described by the great mind, Einstein.

Time to the general audience may seem to fly continuously towards the future. This is exemplified in this commonly used proverb, "time and tide wait for no man." Time is an enigma. Sir Isaac Newton defined time as a constant for everyone at every place. He considered time as an immutable property of the universe. Einstein refuted this idea of Newton and referred to time as a relative entity. In other words, he attributed time to be different for each and every individual. He drew a connection between space and time and no longer considered them as two separate entities but one, space-time. If one is moving, then she/he is moving through space relative to time; and if she/he is stationary, then that person in moving through time. This is based on the fact that passage through space alters time, and this theory has been tested like none other in the field of science.

The effect of space and time cannot be felt much in a limited confinement such as the earth. However, this is not true when talking in terms of light years. Let me start explaining this concept. Life itself is nothing but capturing of individual moments through the passage of time. Individual moments include everything we do with time. A person who is stationary in one corner of the earth, say you, will share the same time as the person 'X' in a planet 15 billion light years away from us. This is because the clock ticks at the same rate. However, let us say, the person 'X' starts to move away from you on earth at an ordinary speed. In that case, the person 'X' no longer shares the same time as you on earth. The time difference may be a fraction of a millionth of a second only, but with the spatial difference of 15 billion light years, the time shared by the person 'X' would be that of your ancestor on earth. The opposite effect is true if the person 'X' starts to move towards you. In such a case, the time shared by the person 'X' would be that of your, perhaps, great grandchild on earth. In other words, the past, present, and the future are all equally true. In a way, the teachings by the age-old religion, Hinduism, is by far, on target: fate is true; and what has to happen will happen. It is just destiny; and every event is already scripted. This seems to be a perfect example of coincidence; it cannot get any better than this. Does that mean Hinduism got it correct ahead of its time? Did Einstein just simply "borrow" a spice from Hinduism's previously curated spice rack, reiterating what was already postulated by the ancient minds? Perhaps so! Believe it or not; theology or science; emotionally or rationally, the conclusion drawn is identical. So then, the ultimate question is that who scripted these events? My plausible answer to this question is TIME. Time is relative; it dictates all events from the birth of this universe to the current minute.

No era is without its share of brilliance that is experienced by their contribution to mankind at different stages of evolution. It is just that we fall short of appreciating earlier contributions. This approach of ridiculing or dismissing the contributions from yesteryears is on many occasions due to sheer ignorance and partly because everyone craves effortless lucidity. The general audience loves to hear pop opposed to classical; watch a movie opposed to floating alone daringly over the deep seas; and only a fraction of the population will attempt to fully understand particle physics while the blissfully ignorant masses use its applications in the form of modern-day gadgets. At this stage, I will let the reader decide if "what has to happen will happen" is a mere coincidence or an ancient truth attested by Einstein.

3

Flu or a Bug?

Of late, apart from the four calendar seasons, there is yet another season referred to as the "flu season." During this time of the year, we routinely hear the phrase "the flu is going around." I was surprised last week when I heard my 6-year-old son mention to me that there is some bug going around in his class and that his best friend was sick. I asked him to define the bug and he said it causes stuffy nose. The word "bug" is used so callously by adults that it has been registered wrongly in the minds of the little ones.

What is this so called "bug?" Is it a termite, a spider, or a bee? Nope, it is neither. Flu is caused by a virus; more specifically, an RNA containing virus referred to as the influenza virus. Influenza virus is so small that it cannot be seen even under a light microscope. The size of an influenza virus ranges between 50 and 120 nm which is 10^6 times smaller than a black pepper flake. Isn't it amazing how such a miniscule pathogen can wreak such havoc on the body? As you probably already know, this viral pathogen can cause fever, cough, sore throat, body aches, fatigue, and runny or stuffy nose.

But, here's the million-dollar question raised by the common man: Why does flu occur year after year during the winter months? There are actually quite a few recorded reasons, and they are as follows: (i) Flu virus survives for longer periods indoors in winter; (ii) The virus may stay airborne for prolonged periods and can thus infect others through inhalation; (iii) Less hours of sunshine in the winter may result in diminished immune activity due to a sharp decline in vitamin D; and (iv) In winter, humans tend to be indoors more and thus have closer contact with others, which makes it easier for the

© Springer Nature Switzerland AG 2019
S. M. Akula, *Essays on Life, Science and Society*,
https://doi.org/10.1007/978-3-030-28775-7_3

flu virus to spread. Honestly, neither of these propositions make any sense due to the following reasons:

(i) The virus is relatively stable outside the human body irrespective of the temperature. It can survive outside the human body effectively for a couple of days as tested on banknotes in a recently published work. Now, there are indeed some differences between virus strains, but overall, they can survive at low pH values (pH = 5.5) as well as at drastic short-term temperature fluctuations (between 40 °F or 100 °F). This defies the dogma as influenza virus is an enveloped one. Scientifically, non-enveloped or naked viruses are stable outside the human body or any host as opposed to an enveloped virus, say, a herpesvirus. This is a fine paradox about the virus and the extra envelope coat that some of them wear. When a virus possesses this outer layer called the envelope, it makes it highly sensitive to the environmental conditions, as the envelope is predominantly made up of lipids that are easily degraded by commonly used disinfectants, temperature changes, and pH variations.

(ii) The ability of the virus to be suspended in air does not change drastically through the year. However, it is worth mentioning that the humidity affects the virus particle size which affects its ability to stay suspended in the air. Also, extreme dry conditions may affect the envelope and reduce the virus's ability to survive in the environment.

(iii) Vitamin D does play a role in regulating the innate and adaptive immune response. However, there has been no correlation deciphered between two-to-three-hour shorter days resulting in a sharp vitamin D loss so as to selectively predispose the person to influenza virus alone. Also, there are regions around the globe where there is not a whole lot of difference between the daylight hours during a calendar year. The best example is Chennai (a port city in southern India) where there is no big difference in the hours of a day during a year. In fact, the joke is that Chennai has three seasons in a year: hot, hotter, and hottest!

(iv) In winter months, people tend to be indoors, but this cannot be the major reason for the spread of flu during the winter months because this concept is biased from the start to finish. It is assuming that the whole world has one type of climatic pattern as observed in the Western hemisphere. This is not true; in fact, the weather during the winter months in Southern India is so pleasant (about 70 °F) that people tend to be out and about even more so than in the summer months.

The scientific reasons for why we tend to succumb to the flu during the winter months versus during the other seasons is primarily because the virus replicates better at lower temperatures coupled with the fact that it is best transmitted at a low humidity of 20%. The viral transmission is almost negligible when humidity reaches about 80%. Flu seasons occur during opposing times of the year in the southern and northern hemisphere. With modern day world travel flu can move quickly with the changing season.

The next obvious question is: Where does the flu virus take shelter during the summer months only to infect us again during the following winter? Does it stay dormant outside the human host? Does it stay dormant inside the human host? Unlike herpesviruses, influenza virus does not cause latent infections; in other words, the virus does not stay dormant inside or outside the host. To understand this, one needs to understand the viral agent in question. Evolutionarily the flu is one of the most highly-diversified pathogens. There are three major types of the virus: influenza virus types A, B, and C. Human influenza virus types A and B cause the disease year after year in the human population. Influenza virus type C generally causes only a minor form of the illness and is relatively rare compared to the other types. This virus exists in multiple types and strains, making it complicated for the human immune system to thwart it. To add to the misery, influenza A viruses are found in many different animals including birds, pigs, horses, whales, and seals. Though, influenza B virus is largely considered to circulate only among the humans, recent scientific work seems to show its ability to infect pigs and maybe other animals. Thus, these other hosts serve as potential natural reservoirs of infection, fueling the virus through to the next season. In nature, any pathogen that has evolved a means to thrive in multiple hosts is a successful pathogen. For example, swine influenza virus, a common pathogen of pigs worldwide and causes flu in them. Transmission of swine influenza virus from pigs to humans is not all that common. However, such a zoonotic transmission is always possible as it happened in India during the early months of 2015. Swine influenza virus strains include influenza A virus (H1N1, H1N2, H2N1, H3N1, H3N2, and H2N3) and influenza C virus. One of the main factors that enable influenza virus to infect multiple species is that it is an RNA containing virus which accumulates mutations 350 times faster than a DNA virus. Accumulation of mutations results in genetic changes critical for a pathogen's ability to adapt for better infection of new hosts. The ecological mechanisms always ensure that the transmission of pathogens between more than 2 different hosts is more than often bound to be asymmetrical. Such asymmetries in interspecies transmission of viruses, including influenza virus is the key to the emergence of viruses that are antigenically different. People have little or poor immunity against such mutation-driven antigenically different viruses resulting in a pandemic. Acquiring mutations is a double-edged sword: on one hand, it allows the pathogens to adapt better to different species while on the other hand may result in a virulent form of the virus.

In a way, humans should thank their lucky stars that such a dynamic and sturdy virus as influenza virus that intimidates the population year after year is not nearly as lethal as the Ebola virus. By December of 2014, I was interviewed by a news channel about the threat of Ebola virus on US soil. In fact, right at that time, all the news channels, as always, were making stories of the Ebola outbreak in the triad nations of Africa: Guinea, Sierra Leone, and Liberia. To the dismay of the interviewer, I laughed it off and mentioned that the worries were in vain and that Ebola would not end the world. I was 100% sure of my judgment and 3 years later, the world is still fine. I based my conclusion on the fact that the number of outbreaks outside these three countries were literally negligible in that year and that the Ebola virus, unlike the influenza virus, really has no proven natural animal host. People concluded bats to be a natural reservoir for Ebola virus and that the virus was transmitted to humans by eating bushmeat of bats. Once again, there is no truth to it. There was not even a single case of Ebola from a neighboring country, Ghana, where at least 100,000 or more bats are consumed every year and it is considered a delicacy. As lethal as Ebola virus may seem, for it to be successfully transmitted around the world over a sustained period, the virus must be well adapted to survive in other hosts as well. With respect to Ebola virus, humans seem to serve as a rather dead-end host. As of now, there are only hypothetical theories as to how the Ebola virus could have been transmitted from animals and none said theories have been effectively proven.

The major problem with treating virus causing diseases of the upper respiratory tract is that there are more than a couple of these "bugs" that can cause diseases of the upper respiratory tract. The viruses that can cause infection of the upper respiratory tract are influenza A and B viruses, adenovirus, coronavirus, rhinovirus, metapneumovirus, respiratory syncytial virus (RSV), human parainfluenza virus (HPIV-1, -2, & -3), and Chikungunya virus. Apart from this list, bacteria referred to as *Bordetella pertussis* and *Mycoplasma pneumonia* can also cause infection of respiratory tracts in humans. When so many agents can potentially cause infection of the upper and lower respiratory tract, resulting in clinical manifestations like the common cold, pharyngitis, laryngitis, tracheitis, bronchitis, bronchiolitis, and bronchopneumonia, it makes diagnosis very complicated and difficult. A friend of mine who is a division chief of the department of infectious diseases mentioned in a lighter vein that if a doctor is not sure of the pathogen, she or he calls it a "bug." Diagnosis is critical only in immune-compromised patients, infants, children under high-risk conditions, and the aged.

Over the course of the modern documented years, science has undergone gradual change in the way it is perceived. The contemplative approach has given way to a more manipulative form. The contemplative approach manifests based on passion for a subject matter while the latter approach manifests based on the desire to gain control or power. In earlier days, science progressed immensely due to the limitless passion that man had for the field. Boundless passion leads to lust or love for everything around us, and in a way, such passion is responsible for all the developments we see in the modern world. Of late, science is being used as only a tool to generate money. Even the approach to fund scientific research is now built on the foundation of networking. The current guardians of science are fully drunk with power and money. The seeking of the truth is quickly and steadily reducing. There is always a court of law for ethics, but no ethics are followed. Science and medicine are important but they pale to insignificance when compared to human bonding, love, and emotion. A scientist must value the human emotions and if his inventions do not in some way benefit the sentiments, it will get disregarded with time. I have deliberately omitted doctors here as their direct role in inventions in biological science is limited, sadly. If this approach continues, the government will fail to make life tolerable.

We humans always miss the crux and put in more efforts on unimportant things. Interestingly, this happens more so willfully by corporate giants and government bodies who care more about churning profit than anything else. Many times, clashes of ideologies impede mankind's progress. Let me give an example where one giant of a personality whom we are all aware put our progress into a retrograde spin. We know of the great American scientist, Thomas Alva Edison. Per the Thomas Edison National Historic Park, Edison held 1093 patents. The first thing that comes to our mind upon the utterance of "Edison" is the bulb and the direct-current (DC) system. However, little do we know of Nikola Tesla, from Serbia, who worked for the famous Edison. Tesla was a genius himself. Among a long list of inventions, Tesla was instrumental in developing the alternating current (AC) technology. The number of patents held by Tesla may seem to be relatively small compared to that of Edison, but far more practical. His contributions to technology include AC, neon lamps, Tesla coil (interesting read if you have time), X-ray, radio (many believe it was invented by Marconi but again, if you have time, read about the controversy surrounding it), remote control, electric motor, robotics, laser, wireless communications, and limitless free energy. Thomas Edison did everything in his power to discredit Tesla's invention of AC to promote his less efficient DC system. To prove his point, Edison went to the extent of even electrocuting animals like elephants using the AC. Edison's histrionics did not end there, as many believe he was instrumental in having bankers opt out of Tesla's projects. There are even suspicions that Edison had a hand in the U.S. patent office's reversal of its original decision to award Tesla the patent for radio; somehow the patent was awarded to Macaroni instead. Today, the local power grids that supply electricity around the globe are based on Tesla's AC. Tesla's idea powers the whole world. Tesla was a genius and worked to solve the problems of the everyday common man, but the shrewd or for that matter, the less intelligent yet wealthier owners of corporate firms always have the upper hand. Tesla died penniless. My friend, Adi (for Adrian) who wrote the preface for this book considers Tesla vs Edison as his favorite example of using money and power to manipulate science and progress. Unfortunately, the situation continues.

On the same note, I would like to talk about this whole idea of vaccinating against this "bug" called flu. Why do we need a vaccine for flu? If you take medicine, it will last 7 days and without medicine it is going to last a week. All these so-called vaccines are made by different private companies in the USA and are approved by the U.S. FDA. It is a well-known fact as admitted by the Centers for Disease Control & Prevention (CDC) itself that it is not possible to predict with certainty which influenza virus will predominate during a given season. Influenza viruses are constantly changing. The strains that are chosen for the Northern hemisphere is the predominant strain circulating in the Southern hemisphere and *vice versa*. They can change from one season to the next and even transform during the same season. Given all these variables, the so-called "experts" must pick the correct combination of the virus well ahead of the start of the "flu season" to manufacture the custom designed vaccine for that particular season. It is like searching for a needle that ain't in the haystack. The act of "experts" picking the correct combination of the virus is like a 6 year old rolling a dice and waiting for it to land exactly on the number 6. The only difference being that the child plays for fun while the so-called 'experts' play with time, money, and hope of poor citizens. There is absolutely no guarantee that the flu vaccination will be effective. Even the companies do not stand by their product. So, this is one of the greatest jokes of the era, along the same lines of how corporate firms played around with Tesla's inventions to fatten their own pockets. Within the last 2 years I had no other alternatives provided but to have the flu vaccination and it has been no different than the several years prior to it when I chose not to be vaccinated. In fact, last year I felt cheated as I had caught "the bug" (flu) a week after my vaccination.

Scientifically speaking, there has been no benefit from this sort of black-magic flu vaccination, and I am convinced it is not going to have any perks in the future. However, have you ever wondered who pays for this no-guarantee vaccine?

In my opinion, you can load the population with either dead or live virus (vaccinate) if it has a guarantee of preventing the disease in the vaccinated individuals without any untoward effects. The best example is poliovirus vaccination which has almost dramatically reduced polio virus-induced paralysis to negligible incidences around the globe. Poliomyelitis is a debilitating disease that can paralyze young children worldwide. Polio can also cause death in about 2–5% children and in up to 15–30% adults. In the US, live attenuated Sabin oral vaccine was administered to effectively bring down the number of polio cases. Live vaccine was being used in the US while the rest of the world was using the killed vaccine. However, we (US) had to change to the killed Salk vaccine in 2000 to avoid what is called Vaccine-Associated Poliomyelitis (VAP). VAP occurs in every 1 out of 2–3 million doses. In other words, every 1 out of 2–3 million children who received this vaccine succumbed to the disease. Morally and ethically, we changed from the use of the live vaccine to the killed version in an effort to make it foolproof, guarantying the individual an effective treatment. In the case of flu, it is a fun ride for companies to generate the ineffective vaccine and benefit immensely out of this sham business. Such vaccination practices can help medical offices make a beautiful pie chart over the years to fodder statistical data analysis about what can be attained. These statistics that describe the outcome of a vaccination program towards a non-lethal disease like the flu are routinely a farce. If the disease is not life-threatening, a common man seldom visits a doctor. I would not be surprised if there is a drive in the near future to make this world rid of flu by vaccination. It is not a far-stretched imagination as it will immensely boost the American economy.

If the government and corporate world are really looking for the magic bullet to combat the flu, vaccination is a mute effort. Vaccination policy will definitely fill in the coffers of the rich corporate governing bodies, but it does not alleviate the situation for the common man. What then is the alternative situation? There are alternative methods to treat the virus, however, to do that, we should understand the biology of influenza virus. It is a segmented RNA virus that accumulates mutations at a high rate. Making vaccinations based on prediction models will fail miserably. It is like shooting in the dark. The restructuring of approach to treat the flu is as follows:

(i) We need to understand that the flu by influenza virus is not a life-threatening disease on the same level as tuberculosis, malaria, HIV, or heart diseases. Let me give you an analogy to explain this fact. Pneumonia is an inflammatory condition of the lung affecting the alveoli and can be life-threatening. Pneumonia can be caused by a variety of pathogens like bacteria (*Haemophilus influenza, Chlamydophila pneumonia, Mycoplasma pnumoniae, Staphylococcus aureus, Moraxella catarrhalis, Legionella pneumophila*, and gram-negative bacilli), viruses (influenza A and B viruses, adenovirus, coronavirus, rhinovirus, metapneumovirus, RSV, HPIV, and Chikungunya virus, cytomegalovirus), fungi, parasites, and idiopathic of nature. Pneumonia (that can be caused by any one of the above long list of pathogens) can cause death almost during each and every period of human life; an exception being between ages 45 to 64 when the death due to this is at a recorded minimum. A death rate of 1.9% (ages 1–9), 0.7% (ages 10–24), 1.0% (ages 25–44), 0% (ages 45–64), 2.4% (ages 65 and over), and 3.1% (ages 85 and over) due to influenza and pneumonia has been reported by national vital statistics reports (dated December 20, 2013).

(ii) We need to cut down millions of dollars being funded to generate vaccines for flu and instead focus on attempting to develop a combination drug regimen targeting the virus entry and replication process.

(iii) We need to develop better management of the influenza virus-induced symptoms.

(iv) We need to mobilize money to effectively provide the infected general population with good and affordable supportive hospital care. Most people who die around the globe are above 65 years of age coupled with poor lifestyle. Honestly, you need a reason to die or else find medical methods to increase the productive lifespan of humans.

Harping on those lines, I strongly believe that the manufacturing of vaccines must be strictly regulated. Currently, it is poorly regulated and is in a pathetic state in the US. The minute a new 'bug' is described, a bunch of scientists whom I often refer to as the vaccine-jockeys start working on developing vaccines. The best example is the severe acute respiratory syndrome (SARS) outbreak in China. It was in November of 2002 when the first outbreak of SARS was reported in the Guangdong province of China. It was immediately declared a global disease; though the number of cases was miniscule compared to any disease we know. Immediately, labs in the US started working on developing a vaccine to this so-called deadly (mockingly) disease. There is nothing wrong with pouring in ideas to develop a vaccine against a potential threat but things should always follow a rationale. The following questions must be addressed before actually proceeding to invest money in developing vaccines:

(i) What is the etiology for the said disease?

(ii) What is the impact of the pathogen on humans?

(iii) Is the disease zoonotic?

(iv) What is the lifecycle of the pathogen?

(v) What is the epidemiology of the disease?

(vi) Has the pathogen been sequenced?

(vii) Has the structural and non-structural proteins of the pathogen been thoroughly analyzed? and

(viii) What is/are the protein(s) in the pathogen that is immunogenic or more importantly neutralizing?

(ix) The most important of them all: Vaccines are meant for preventing a disease. But, as seen in the case of the flu, even after vaccination if there are approximately 31.4 million outpatient visits, 200,000 hospitalization, and about 5000 deaths, the vaccine is just a futile attempt. It is just a way to boost the revenue for the healthcare division at the behest of the poor citizens. Flu vaccines have not changed the scenario at all. It is essentially a placebo to the patients and a money maker for the doctors. As in many cases, the highly worshipped vaccine pushing doctors are either just ignorant or they pretend to be believers. I can tell you with 100% honesty that I have had such discussions with scientists who are directly involved in the design and development of this and other vaccines, and admittedly they all know the flu vaccine is not a magic bullet as in the case of the poliovirus vaccine.

It is crucial to understand the biology of a pathogen. What is the purpose or rationale behind initiating the development of a vaccine before you understand the pathogen that its being designed to rid? That is like going to a war without any strategy. Can you imagine that the first scientific paper on vaccine for SARS was published only a year after the outbreak? With such little time lapse, not a whole lot was known about the pathogen itself. Goodness gracious! This pathogen turned out to be a dummy or a pseudo-pathogen, at best. All the panic and mass media coverage for less than 1000 deaths in a handful of countries to date. I am sure that my point-of-view regarding the matter is not music to the ears of vaccine developers. Believe me, I am not being heartless but rather trying to present my case against the commonly peddled flu vaccine. There are so many other pathogens that are not only debilitating but can cause huge loss in terms of mortality year after year. I am NOT against treating flu, but I am advocating scientists to pour money into developing alternate methods of treatment as opposed to holding on to a preventive approach of vaccination that has no guarantee. Why waste money on a burger gone bad?

4

Why I Chose Evolution Over Religion!

Both evolution and theology are awesome ideologies! They both describe the origin of life in their own interesting way. Of course, religion is a more popular concept than evolution, as most of us are born into one of the religions in the world (no choice there) and most of us continue to blindly follow it as we journey through life. Also, it is a fundamental law for the religions to "pooh-pooh" the concept of evolution. Though there are several differences between the two ideologies, I would like to list just two of them that I deem most startling:

1. Apart from the origin of life, religions also do the job of advocating rules. Basically, they serve as judges to settle all day-to-day problems like rape, blasphemy, cheating, infidelity, and etcetera with rusty outdated codes of conduct.
2. Modern faiths have clear and perfect (assumed) solutions to all the questions in the world. They have foolproof answers to topics from health to global warming in contrast to the ever-doubting scientists toiling their minds to increase life expectancy and identifying newer concepts to make life better on this planet.

I was born into a Hindu family in a small town about 10 miles south-east of a reasonably famous city called Trichy in the southern state of Tamil Nadu, India. At home, I was taught "good" ethics and values based on Hinduism. I was conditioned into believing that prayer makes a difference in life. According to the ideology, Gods and Goddesses are watching every move that we make, and it is imperative that one makes no mistakes to avoid their

© Springer Nature Switzerland AG 2019
S. M. Akula, *Essays on Life, Science and Society*,
https://doi.org/10.1007/978-3-030-28775-7_4

wrath. At my school which was run by a Catholic organization, I met friends from many different religions each believing in their respective savior. The following incident that I am recollecting took place when I was about 12 years of age. It was customary for most students to visit their respective place of worship prior to exams to get blessings from "The Almighty" to score good grades. I had never done that, as I prayed to God in my home. Honestly, I was not an honor roll scholar in my school days. I was just an average Joe. Out of curiosity, one day I decided to pray as "I want to ace my exams," prayer like the rest of my peers. I thought that by seeking the blessings of the almighty, I could improve my standing in the class. I visited the temple before exams every day for a whole week only to find out that my "I want to ace my exams" prayer had no effect on my rank. My rationale to visit temple every morning before exams might have been quixotic but I quickly realized "God helps only those who help themselves." It dawned upon me that if I could help myself and be pragmatic, then I did not need God's help. At this point, I lost 25% of belief in any form of religion. As a young boy, I distinctly remember having arguments on religion and its beliefs with my friend Ramya. Ramya was in the same class as me, but unlike myself, she just so happened to be regarded as one of the most intelligent students in my school. Let me not fail to mention that she was from a very religious family and thus, so was she. Once I went to the veterinary school, I lost touch with Ramya who went on to study medicine.

During my school days, evolution was never taught in the classroom in a detailed manner. In fact, I don't think the situation has changed much from then till now, and that is the reason why evolution is one of the most poorly understood phenomenon's even among the educated mass. A sad reality, indeed. So, why are talks of evolution avoided like the plague? Well, this is partly because of the daunting and taunting influence of theology. When the word 'evolution' is heard the only thing that comes to mind of a common man is Charles Darwin and his famous thought: 'the fittest survive'. Interestingly, this thought is only the tip of the evolutionary iceberg. I am going to attempt to simplify the concept of evolution that has shaped this world with innumerable life forms over millions of years irrespective of the religion that was practiced in different corners of this pristine Earth.

As a professor at the medical school, apart from teaching, I also actively conduct research. I have encountered at least once in my own life as a scientist an instance when my own idea was supported by work from across the Atlantic Ocean with someone whom I had absolutely no links and got published in the same journal and volume. I was always baffled by this fact. But not any-more, as I read and understood how the modern theory of evolution unfolded.

There were two great minds that lived at the same time and place (England) who are responsible for proposing the famous and most popular theory of evolution based on natural selection. They were Alfred Wallace and Charles Darwin. Wallace believed strongly that species changed over time. His work on genesis of species titled "On the law which has regulated the introduction of species" was published ahead of Darwin's. He was the first to propose the fact that the fittest survive based on his field investigations in Singapore and the Indonesian islands that included Borneo, Bali, and Lombok. In fact, his article was submitted when he was working in the deep jungles of Borneo. He concluded that life is just a struggle to survive the odds in nature. Those that survive are better adapted. Useful variations of a species tend to increase while the not-so-useful variations are lost over time. Wallace was a humble self-made man who reasoned the crux of evolution while simply trying to make a living by collecting specimens from different parts of the world. Wallace's work put immense pressure on Charles Darwin who was taken aback by such perceptive published work from the lesser known bee collector, Wallace.

Wallace, on the other hand, had such a colossal regard for Darwin that he sent his next work on the "Natural selection" to Darwin himself from the Indonesian islands. That was the final blow to Darwin's aspiration to be the first to propose the theory of evolution based on the mechanism of natural selection. Though perturbed, Darwin directed the article to the Linnean Society for publication. In the society, it was decided that the work would be published as one manuscript titled "On the tendency of species to form varieties; and on the perpetuation of varieties and species by natural means of selection" co-authored by Charles Darwin, F.R.S., F.L.S., & F.G.S., and Alfred Wallace. This story is an epic by itself. Here is a well-educated and qualified Charles Darwin who had worked for over 20 years to formulate his natural selection theory. Then seemingly out of nowhere, Darwin, the scholar, was "given a run for his money" by Wallace, the humble field worker who simply collected beetles and butterflies to make a living. This to me is a defining moment in itself, as it conveys the reality that education does qualify a person with laurels but all who have a desire are equally equipped to adequately rationalize irrespective of their credentials.

To gain an appreciation for evolution, you must be willing to do the same. One does not need to be an atheist to understand the science of evolution but rather be willing to read and analyze the facts in support of it. Evolution is like the "Theory of relativity" in physics. Every high school educated person knows that Einstein proposed the theory but only a few attempt to gain an understanding. The reason being is that comprehensively reading such a theory is not as simple as reading a novel or memorizing the 206 different bones in the adult body. Delving into understanding Einstein's works may take multiple readings (actually studying) to get an inkling of an idea of what was being conveyed by the great mind. Along the same lines, understanding evolution may require time to read, understand, assimilate, and then amalgamate a model in your mind on what was proposed almost 158 years ago. What a pity! Interestingly, there lies the difference between understanding evolution and religion. To understand evolution, you need to have a clear appreciation for the biology of life, biochemistry, molecular biology, and at least a little bit of physics with willingness and interest to question and think. On the contrary, with religion, most don't have a choice and the ideology is thrust upon them at birth. Very few individuals over the course of time decide to shift their beliefs for personal gains. More importantly, there is no chance of questioning the authority of religion; at least that is 100% true in the case of modern religions.

The main points of evolution as proposed by both Wallace and Darwin are,

(i) Variations are commonly observed between the living beings of the same species. The best example to explain this phenomenon is that, in general, siblings do not resemble each other. There is always bound to be variations even among children born to the same parents.

(ii) More living beings of the same species are born than could survive. This is where they believed the fittest only survive.

(iii) Living beings of the same species with better and favored characteristics to endure and reproduce have a better chance at survival.

(iv) These favored characteristics are passed on to the next generations. However, at the time when this was proposed, they did not know how the characters could be inherited from one generation to the next. This was because knowledge in regard to genes was limited then. In a way, the final point made by Wallace and Darwin was in fact an extension of the theory proposed earlier by the French naturalist, Jeanne Baptiste Lamarck, in 1801. Lamarck strongly believed that traits important for the survival of an organism were passed on to its next generation. On the same lines, traits or organs not critical for the organisms' survival were lost. As one can see, the idea of natural selection took shape over a period of time, even prior to the works of Lamarck.

But before I go any further, one must be wondering why Darwin's name is so prominently connected with the modern theory of evolution as opposed to anyone else. Has the world forgotten about the worthy works of Wallace? I have wondered about that too. Politics has had a role at all levels and at all times; science is no exception. Darwin was well placed in the Victorian upper class of society. He was from a wealthy family and coddled away from life's pressures. He was educated in Cambridge. Contrarily, Wallace was from a dwindling lower middle-class family with limited resources. Unable to afford a sustained education, Wallace did manage to attend the Mechanics Institute of London. While Darwin's ride on the Beagle across South America itself was self-funded. The richer man (here you may call him the fittest in terms of evolution) was better accepted in the society while Mr. Wallace, the poorer man with equally as remarkable contributions, has seemingly been erased from the modern theory of evolutionary discussion altogether. This is yet another perfect example of how our Earth has evolved over time. The fittest (in this case, the richest) survived in this instance too!

During the 1800s when the whole world was strongly influenced by religious beliefs (not much has changed since) on creation of life, the divergent thought proposed by Darwin and Wallace was mind blowing. Even after 200 years, it is highly impossible for an educated man to even think on the lines of evolution. In general, a butterfly is looked at in surprise for its extravagant array of colors; a tortoise is looked at in awe for its unique structure and shape; a beetle is destroyed with our foot at first sight; sight of a chameleon means a not-so-safe place. On the contrary, all these very same observations drove the unique minds of Wallace and Darwin to work out a theory on natural selection. Mind you, back then there was absolutely no clue on the role of genes in transmission of traits from one generation to the next. Therefore, logically it was concluded that there was constant interaction between the species and its environment and that it was nature (and not God) that was directing the selection process which resulted in new species. Whether one believes in it or not, evolution has shaped the life on Earth for over millions of years. There is plenty of evidence to support it and I am not going to discuss it in this essay. However, the concept of evolution as proposed in 1800s is just the beginning or rudimentary, and to some extent, modern science can explain evolution on a completely different plateau. Does life evolve and did evolution shape all life in the world you see today? Absolutely! But modern-day genetics will teach us that life evolved not on the directions of nature as proposed by the Wallace and Darwin but by the genes. In a way, the concept of evolution as proposed by Wallace and Darwin cannot be taken as the ultimate truth. It has evolved with new findings based on scientific research. Even about 5 years ago it was believed that genes are the finest individual machines that could express a particular protein. However, the latest scientific research takes things a step further and reveals that these genes are in fact regulated by another set of regulators called micro RNA (miRNA). On those same lines, evolution itself, a field of science, also has evolved over time.

What is life in terms of a biologist? It has an extended history. I would define it as the time when life in any form came into existence on this planet. The modern religions claim life to have originated from Adam and Eve. A few religions (due to their short comings) do not bother to date this time period, while Christianity dates it to be precisely the night preceding Sunday, 23 October 4004 BC. This date was estimated by the Archbishop of Armagh, James Ussher (1581–1656). But modern science proves this estimate to be wrong. Based on fossil records, scientists date the origin of life on Earth to be around 3.5 billion years ago when unicellular life came into existence. It sounds rather funny as modern religions stake their claim over creation of life. The paradox is that man came to know of these religions only about 4000–4500 years ago. Interestingly, while science and understanding of the powers of nature have steadily been changing, man has stuck to his age old and rusty ideology. Now, let us get pragmatic. Worshipping supernatural powers has existed for quite some time; even before the new religions came into existence. It is just a belief and more so a tool that gives man his confidence. In the initial years, man worshipped the natural forces which intimidated him. Over time, beliefs slowly gave rise to organized religions; different religions were practiced in different corners of this planet while life persisted without any partiality irrespective of the belief. None of these religions have had an effect on the numbers of crimes committed. If there is secularity, it is obvious in the food consumed, crimes committed, and the sexual intercourse irrespective of what religion one practices. This is a clear hint that there is no supreme belief or religion that can control the behavior of man. I do not want to take anything away from the religions as they all preach only one thing and that is kindness. Religions are a must for weaker minds to get the much-needed mental support and on the other hand for idle man's brain to fight over and settle scores. When I realized this anomaly, my faith in religion dipped to 50%.

I have had many discussions with myself and at times with friends of mine regarding why I chose "evolution over religion". The best discussion I had was with Ramya again and this was at the Dallas airport as I was waiting to board my plane back to Kansas City. I was on my way back home after having my second interview with the folks at the UT South Western University the day before, but my flight was delayed due to icy conditions (perhaps, a divine intervention). I intimated the delay to my wife and as I was getting settled with my laptop, I was pleasantly surprised to see my childhood friend Ramya again. This time around, she was on her way to attend a meeting on how life forms respond to fluctuations in gravity. During our conversation, I came to know of how she got interested in physics while attending medical school.

The discussions I had with her had a deafening impact on me. When we do not understand a natural phenomenon, we call it divine intervention. We accept it as God's work because that is the easy route for a mind that is not willing to search for the truth. Ramya gave a perfect example of how science and religion shaped the modern world. In her opinion, religion has served as a speed-bump to the progress of science. Here is one perfect example. One of the most fascinating fields of science is physics—which can at times be difficult to comprehend by the general audience. Unlike medicine, the advancement of this field of science requires not only intelligence but quite a bit of imagination and reasoning beyond commonsense. Earth in general was considered to be flat by all the ancient cultures as dictated by their religious beliefs. Religions, rarely dealt on difficult topics such as those described below. It took a lot of time to conclude the Earth as a spherical object. Aristotle (385–322 BC) proposed a geocentric model with Earth at its center while planets, the moon, and stars (comprising of heaven) revolved in perfect circular orbits around it. It was Alexandrian mathematician, Ptolemy (90–168 AD) who further perfected Aristotle's model keeping Earth at the center of the universe. By conducting systematic observations, Ptolemy prescribed a more sophisticated model to account for the retrograde motion and the differences in the brightness. This theory was approved by the church as the Bible suggested the sun to be in constant motion while the Earth was stationary. The geocentric theory strongly believed that the Earth was stationary because no one actually felt the Earth moving under their feet. This theory might have been ridiculed by many (not all) in present times but it was a fine beginning. Believe it or not, it was Ptolemy's map that was being used to navigate around the world till Columbus (1451–1506) came into the picture and such was his influence. It took over a thousand years before Nicolaus Copernicus (1473–1543) proposed the heliocentric model of the universe with the Sun at the center. He delayed publishing this work for quite a bit of time primarily to avoid any form of reprimand by the church. Finally, when it was published, the book was dedicated to Pope Paul III (Only the Pope and God knows how and why the masterpiece was dedicated to him). Copernicus was street smart. Unfortunately, his successor, Galileo Galilei (1564–1642) was not. He argued with the church that the Bible may not have solutions to every natural phenomenon while supporting Copernicus's heliocentric model. This attitude was considered blasphemous by the clergy. He was forced to recant his belief that 'Earth moves around the sun' and while doing so, he uttered the most famous phrase "And yet it moves." For his act, he was kept under house arrest for the rest of his life. Over the years, our knowledge of the Earth, the universe, and our nature kept growing steadily but slowly. Then came the genius of a mind called Sir Isaac Newton (1642–1726). He discovered the laws of motion, gravitational force, and many more even before he turned 26 years of age. Sir Newton had no reference to God when he solved the puzzles related to motion, gravitational pull, or even the two-body problems because he could solve it. However, when it became too complicated to him while addressing the manner by which the 6 planets (now 8), the moon, and the asteroids moved around the sun, he immediately invoked it as the act of God. This is not all, even the great Albert Einstein faltered when he introduced God in his rational. Even though he wrote the first manuscript on quantum mechanics, he could not believe in quantum entanglement where two particles at a distance according to him had a spooky action. He reasoned it as "God does not play with dice with the universe." But we know that quantum entanglement is reality. When man can reason it, he calls it science; but when his faculty of senses fails to appreciate a phenomenon, he terms it to be an act of God. In Ramya's own words: "as you can see, over the years science has clarified every single miracle or myth into a reality." At the end of this discussion, I had lost complete faith in all these manmade and dictated religions. This along with the unparalleled sorrow and atrocities I saw around me, I completely lost faith in the almighty. If, I, as a normal human being with no extraordinary powers cannot put up with disparity, why did God, a super natural power, create this world littered with disparity from pole to pole? I saw poor children begging for alms; infants dying; children born blind, deaf, and mentally challenged; God-fearing people never bothered by breaking rules; and the list goes on. This world is not for the peace loving, unselfish, innocent, and the non-defensive clan. Such calm, non-destructive, and helpless people are continuously being wiped off the surface of this planet. Only the tough, shrewd, vocal, and the atrocious survive and make it; in other words, the fittest survive. Well, on this one aspect the modern day religions agree by systematically exploiting to proselytize the poor and powerless minds. Religions fail to understand that the majority does not mean a victory or the truth; it just reveals how insecure they are.

Even more so, our (mankind's) perspective on religion itself has undergone a slow but steady change over time. Old perceptions are slowly giving way to the new ones. In fact, religion itself can't even avoid evolution. Religion might be as big a deal in the minds of people today as it has been at any point in time, but religion pales into insignificance as compared to life, time, and space. Life has survived in this definite space of Earth for as long as 3.5 billion years of time with changing beliefs and faiths over time. After all, man introduced these new age religions only in the last 4500 years or so. At this pace, one can argue to see a new dimension in his belief in the next 1000–4000 years to come.

As a rationalist, I still had one question lingering in my mind. Who created this beautiful Earth, life, and the limitless universe? It cannot be an act of one man, lady, or God; why then would you have millions of uninhabited bodies in this vast universe? It must have been a waste of time to serve a moot point (unless the religious thoughts have a clue; I honestly doubt it as they have always been written or passed on by mortal humans with lack of acumen in math, physics, and chemistry). What is the origin of life? It is nothing but a chance biochemical event under optimal conditions. Who created the universe and how was it created? This is work in progress. If you had asked me when I was a 10 year old kid, I would have said atoms made up the universe and as I grew older, my ideas evolved based on what I learned at school and more so from reading on my own. It changed from molecules to atoms to protons, electrons, and neutrons to the elementary boson and fermion particles. My interpretation completely depended on my evolving knowledge on particle physics. New things are being unraveled at quite a rapid speed and I am sure, in the near future, science will reveal if the boson field created this universe made of elementary particles or if we are nothing more than just a holographic expression of the information stored in different corners of the black holes. I for that matter, don't want to believe in one creator heading the man-made religion, as I feel I am far superior than these folks who lived thousands of years ago.

5

It Is mi(y) RNA!

I love Einstein for his contribution to science and humanity. The story of general theory of relativity is an epic in itself. More so, on the lines of Ramayana and Mahabharata; with as many twists and turns. Conceivably, great ideas mean greater resistance and pressure from all sides. I get goose-bumps when I read about Einstein's' contribution to modern day life. It does not stop at just $E = mc^2$. His contributions are not limited to GPS, smart phones, and the numerous applications of the laser. I doubt if the modern-day physicians really have any appreciation for his contributions to medicine. All his revolutionary ideas were as a result of intense competition among his peers. An example of which was the 8 year long struggle to come up with the mathematical equation to support his theory that space is indeed curved. He had a tough competitor in the German mathematician David Hilbert. Though Hilbert also published his field equations in support of general relativity almost at the same time as Einstein, there was no animosity between these two individuals and Hilbert acknowledged the fact that it was Einstein's theory. This is a result of great minds. Einstein's theories set a new trend and in fact questioned 250-year-old Sir, Isaac Newton's theory on gravity. It revolutionized physics at all levels. Biology has also had its share of new ideas and I am going to talk about one such thing and how it is slowly changing the fundamental concept of how the cell functions. I am careful in using the word 'SLOWLY' as even scientists have not realized the concept in its entirety. As I said earlier, ignorance is the greatest barricade for progress.

For too long, it has been taught that one gene encodes for a protein and that the genes are referred to as the regulators of all the cellular functions in the body. What are genes? They are made up of deoxyribonucleic acid (DNA).

© Springer Nature Switzerland AG 2019
S. M. Akula, *Essays on Life, Science and Society*,
https://doi.org/10.1007/978-3-030-28775-7_5

They are highly stable in nature because of their double helical structure and the hydrogen bonds between the bases. They encode data to generate different proteins in the body.

The genome is made of DNA. The smallest genome of a living creature belongs to a bacterium called Carsonella rudii, and is 159,662 base pairs while marbled lungfish contains the largest genome (133 billion base pairs). The human genome is only about 3000 megabase pairs compared to the one in the marbled lungfish. It makes me wonder how faulty the evolution has been; primarily because it has been dictated by the cruel laws of nature and not the hand(s) of God(s). If it was a design by a superhuman, *Homo sapiens* would have ended up with the largest genome and not the other way around. The human genome is packaged in 23 pairs of chromosomes within the nucleus and each one of these chromosomes contain hundreds of genes. The human genome contains 20,000 genes and each one of these genes encode a single distinct protein. Interestingly, only about 1% of the genome actually codes for a gene and are commonly referred to as the coding DNA or exons. The rest of the genome are generally called the non-coding DNA, introns, or junk DNA. Generally speaking, most, if not all the genes <0.55 kb in length do not possess any introns and that genes ≥0.55 kb do possess introns. Such an arrangement has an important implication in the origin of genes because of one or both these reasons: (i) the original form of a functional primordial gene was ≤0.55 kb in length; and/or (ii) introns were acquired by the higher order life forms during the course of evolution. The non-coding DNA were for a long time regarded as a portion of DNA used to produce non-coding RNA components like different forms of RNA. Many a times, it was regarded as a vestigial product of evolution.

The beauty with science is that active research constantly develops and provides new dimension(s) to already existing concepts and ideas. It is constantly evolving with time. Research in the last decade has generated proof to believe the non-coding DNA segment as no longer a string of junk. It contains crucial elements that regulate the coding gene sequences. In fact, there was an interesting article in the 30th of January, 2018 edition of Scientific American that explains how these so-called junk DNA may be the key to how the brain develops. More importantly, these non-coding sequences are the source for microRNAs. This is yet another proof to demonstrate why it is good to believe in science and rationalize events around you; primarily because it evolves with time rather than being stagnant for eons.

The DNA is so stable that it can lie dormant in fossils for thousands of years. The DNA dogma clearly paints genetics to follow the laws of utopia. Disarray follows an order; the orderly DNA is transcribed into an unstable intermediary, RNA, which eventually gets translated into a protein. RNA is considered unstable because it is highly susceptible to hydrolysis and thus can't survive for long in nature. In a way, DNA has always played a role of a big brother to RNA; partly because of the quintessential role it has in the field of forensic science and archeology. To a great extent, tracing human DNA has been able to reconstruct our past. Apart from this, working with RNA compared to DNA in the laboratories can be troublesome due to the following reasons:

- DNA can be stored at room temperature.
- RNA must be store at −80 °C only.
- Special precautions need to be taken while handling RNA to prevent its degradation.

DNA contains deoxyribose while RNA contains ribose that is characterized by the presence of a 2′-hydroxyl group on the pentose ring. This hydroxyl group in RNA makes RNA less stable than DNA because it is more susceptible to hydrolysis and thus degradation. My laboratory works on DNA viruses. In my own experience, it took me a humongous effort to convert my laboratory to start working on RNA. For example, we could no longer use autoclaved double deionized water. We had to use DEPC treated double deionized and autoclaved water, instead; and the work area had to kept RNase free by cleaning it with RNase cleaning agents. Simple things like these will add up to give a new dimension to the standard operating protocols. Up until recently, RNA had three important attributes in biology and they are:

(i) RNA primordial soup is believed to have powered the origin of life in this planet, Earth. In reality, a combination of small self-replicating RNA in a conducive environment rich in iron disulfide and iron-nickel sulfide are the essential elements crucial to the origin of life in this planet;

(ii) RNA is the intermediate step between the gene (DNA code) and the effector protein it encodes. There is a perplexity in this design because of the fact that RNA actually preceded DNA in the evolutionary cascade. It has been determined that the ribonucleotides were available prior to the evolution/development of the DNA synthesis mechanism. The clue lies in the ubiquitous occurrence of ribonucleotide reductase (RNR); a class of protein critical to catalyzing the synthesis of four deoxyribonucleotides required for DNA synthesis and repair. Hence, RNR is essential to the transition of DNA from RNA. In simple terms, RNA supplies deoxyribonucleotide precursors for DNA synthesis; and

(iii) RNA is monitored to understand the functionality of a gene.

So, what is the function of RNA? To a layman, RNA can be explained as the middleman between the DNA coded message in the gene and the functional protein crucial to triggering each and every signaling in a living being. RNA acts as a middleman because it is nothing but the message encoded by the DNA in a language that the protein synthesizing machine, the ribosomes, understand. One cannot shrug off the role of RNA as the middleman in this process of triggering a gene-related response. In reality, the middlemen are introduced in day-to-day life as in business modules; wherein, they play a role in shortlisting applicants, interviewing, hiring, scrutinizing the performance, laundering money, and so forth. Almost always, the whole process involving middlemen is a less efficient process strictly benefitting only the middlemen for no avail. These business modules will work with greater efficiency if this player of a middleman is eliminated. However, RNA, the middleman has a crucial role to play in the biology of life. DNA may well carry the code for life but if not delivered in the appropriate manner to the ribosome, specific proteins will not be synthesized.

In general, total RNA is the set of RNA molecules within a given cell; and this includes messenger RNA (mRNA), transfer RNA (tRNA), and ribosomal RNA (rRNA). Truly, rRNA makes up the bulk of RNA in a given cell. Each and every RNA species has a specific role in the biology of life and they are as follows:

(i) mRNA: The coding DNA sequence corresponding to a particular gene is transcribed into an RNA sequence which is finally translated to a quaternary structured protein. This RNA sequence that is translated into a protein is referred to as the messenger RNA or in short mRNA.

(ii) tRNA: The role of which is to decode a mRNA sequence into a protein. Each type of amino acid has its own specific tRNA, which binds and carries it to the growing end of a polypeptide chain.

(iii) rRNA: They form the ribosome machinery that includes both the large and small subunits. rRNA molecules not only interacts with tRNA and other accessory molecules but also physically move along an mRNA molecule to catalyze the assembly of amino acids into protein.

All of the above species of RNA molecules have a crucial role in the translation process; in other words, making up of proteins. The first research study describing a direct link between RNA and protein synthesis was published in 1955 by Laird et al., from the University of Wisconsin, USA. However, it was not until 1975 that scientists determined a crucial role for mRNA, tRNA, and rRNA in the protein synthesis process.

Recent studies determined the existence of other known RNA species in the living cells. These were classified as long non-coding RNA (LncRNA) and small non-coding RNA (ncRNA) molecules. LncRNA and small ncRNAs has gained prominence only since early 2000. As the name suggests LncRNA are longer transcripts (>200 nucleotides) and they appear in the order of tens of thousands in mammals with lesser known biological significance.

The small ncRNAs are those that are shorter than 200 nucleotides in length. There are several kinds of these varieties and they are as follows:

(i)   microRNA (miRNA) – discussion in detail to follow below.

(ii) Piwi-interacting RNA (piRNA) – They are a germ-line specific small RNAs that are processed from a long single-stranded precursor transcript.

(iii) small interfering RNA (siRNA) – They may also be referred to as the short interfering RNAs. siRNA has gained importance as a potential therapeutic reagent due to its ability to inhibit specific genes in many genetic diseases. Despite this potential, its application in clinical settings is still limited partly due to the lack of efficient delivery systems.

(iv) small nucleolar RNA (snoRNA) – They are 60–300 nucleotides long non-coding RNAs in length that accumulate in the nucleolus of a cell. They have an important role in the synthesis of ribosomes.

(v)   tRNA-derived small RNA (tsRNA) – This is a class of small ncRNAs generated during the course of maturation of tRNAs. Recent studies have demonstrated tsRNAs to be dysregulated (specifically down-regulated) in a variety of cancers including leukemia and lung cancers.

(vi) small rDNA-derived RNA (srRNA) – This class of RNAs are yet to investigated or lesser known ones. They are supposed to be involved in various cell signaling pathways.

(vii) small nuclear RNA (snRNA) – They play important roles in splicing of introns (segments of DNA or RNA that do not encode a protein) from primary sequence.

Of these small ncRNAs mentioned above, the focus of this essay is going to be on miRNAs. It is easy for people to get confused between mRNA and microRNA (miRNA). There is a world of difference between mRNA and miRNA. miRNAs are encoded by the non-coding DNA sequence. They are usually small, containing about 22 nucleotides and are a part of the small RNA family. The miRNAs came to prominence in early 2000. miRNAs show great diversity in sequence and expression patterns and are evolutionarily widespread, suggesting a key role in gene regulation. The components and the biogenetic pathways associated with the miRNAs are conserved between plants and animals. Among mammals, miRNA coding sequences within the introns are estimated to account for about 1% of the genome, and yet greater than 60% of protein coding genes are regulated by miRNAs. miRNAs control gene expression and regulate a wide array of biological processes by targeting mRNAs and inducing translational repression or RNA degradation.

My lab started working on miRNAs in the year 2016. It is a fascinating field of study and. Can you imagine the following?

(i) A 22-long oligo in the form of a miRNA has been able to paint a new dimension to the otherwise, junk sequences of DNA.

(ii) One miRNA may regulate tens to hundreds of genes located at different loci.

(iii) miRNAs have evolved through ages to an extent that certain viral pathogens like herpesviruses, encode their own miRNAs.

(iv) miRNAs also dictate the outcome of disease conditions like cancers, diabetes, heart diseases, and so forth.

I have always wondered how the biology of life works. Life in all its superiority is held in place by the stable DNA which in turn is regulated by the most unstable RNA. The law of increasing entropy does not apply to an open system as life. Then, is there a reason for such a design of regulation? The answer is yes and it is on the following lines:

(i) DNA is the quintessential of life and its environment. A reasonably good biologist when presented with the DNA of an animal or plant would not only be able to reconstruct the specific living being but also its environment. The complex life has been held intact by this string of three letter codes. The integrity of life has been passed on for over thousands of years while gently accumulating minor changes along the line to drive the evolution of a particular species based on its survival requirements. Such a sturdy task can only be accomplished by a stable, robust, and an effective element such as DNA. DNA can survive the harshest of environments for thousands of years.

(ii) The code for life is contained in the DNA but it is not all as once conceived; as the expression of DNA containing genes is strictly under the control of the highly unstable RNA, referred to as the miRNA. This is a clever design by nature as for the switch (miRNA) to be able to perform its function, it must be in a conducive environment at its optimal functioning ability. In a way, miRNA functions as a trip switch in an electric circuit. It regulates the expression of the genes and at the same time plays as a checkpoint when the conditions are not conducive.

(iii) If I may extrapolate the role of miRNAs in the biology of life, I would conservatively state that they have a strong role in the evolution of species more than the DNA itself. The rationale to this conclusion is the fact that environmental changes directly influence the miRNA expression which in turn regulate the expression of specific genes. The physiological and the pathological events within a living thing is dictated as a measure of miRNA activity. Specific RNA molecules work in tandem as effectors (miRNAs) and the substrate (mRNA) to the DNA that is crucial to sustaining life in every form on this planet Earth.

(iv) Epigenetics factors shape life of all living species including human beings. Epigenetics refer to the changes that affect the expression and activity of genes. These heritable phenotypic changes do not involve mutations to the genes. The factors may be as simple as smoking, drinking and more complicated as DNA modifications like DNA methylation, histone methylation, and histone deacetylation. In short, behavioral activities such as but not limited to stress, smoking, drinking, obesity, and addiction to drugs can all influence miRNA expression and thus the regulation of genes.

There are two other RNAs that people get confused with and they are siRNA and miRNA. Both have a crucial role to play in RNA interference (RNAi). siRNA is a synthetic molecule, exogenously introduced to interfere with RNA processing. siRNA is double-stranded short RNA that binds perfectly with its target mRNA. Also, the siRNAs induce post transcriptional gene silencing. On the contrary, miRNAs are single stranded RNA encoded by the non-coding sequences. Their binding to the target mRNA varies to a great extent: may be perfect or imperfect. miRNA typically silence a gene function by repressing translation. Both siRNA and miRNA gained prominence as RNA-based therapeutics to treat a variety of disease conditions like cancers, diabetes, cardiac diseases, viral infections, and ocular hypertension. Personally, my choice is miRNA-based RNA interference over the siRNA-based tool for the following reasons: (i) miRNA is naturally occurring compared to the artificially synthesized double stranded siRNAs; and (ii) A specific group of miRNAs termed as the epi-miRNAs can directly target effectors of the epigenetic machinery crucial to development of different kinds of pathologies including cancers.

In general, siRNAs are better understood by scientists than miRNAs. This is partly because it has been used as a tool to silence expression of genes *in vitro* in many labs around the globe. I started using it as early as 2003 and continue to use it as it is easy to get it synthesized, cheap, and reliable. Over the last decade, improvements have been introduced to increase an efficient delivery and increase the half-life of it in vitro and *in vivo*. To my knowledge, most of the labs in the USA must have used it at least once by now. On the other hand, miRNA is a relatively new topic and I'm sure currently, many scientists fall short of understanding it. The technique used to identify novel miRNAs is a laborious process that involves the use of relatively new procedures. Also, authenticating a novel miRNA is not only tedious but may also end up on the lines of searching for a needle in a haystack.

The advent of miRNAs in the biology of life will have a compelling effect on multiple things as follows:

(i) In a nutshell, RNA is the unsung hero in the biology of life. If boson is the God particle when it comes to the universe, I would call RNA the creator of life on this planet. Evolutionists strongly believed that life originated about 4 billion years ago from RNA. But to form RNA, both purines and the pyrimidines must be available simultaneously. Stairs et al. (2017) [Nature Communications 8:15270] have recently provided the mechanism how RNA could have formed in the prebiotic world. By this way, RNA (and not DNA) actually carried all the information crucial to life. We share a common ancestor with the primordial RNA; that perhaps carried all the inequalities noticed in the modern world. In a way, it is mi(my) RNA that has been responsible in shaping this world. Small RNAs of a species also plays a crucial role in responding to the environment in such a way as to regulate genes for survival. It is becoming more apparent on how miRNAs including other regulating small RNAs are the major reason for the extinction of animals like dinosaurs, or acquiring new characteristics over time. It is sad, that this knowledge about RNA, the creator of life, is not getting passed on as effectively as the other unimportant materials. If given its due share, knowledge on miRNAs will help us redefine evolution. After all, our understanding of evolution has evolved over ages; just as Wallace and Darwin had no understanding of genes when they proposed their theories of natural selection in 1859. Their contemporary, Gregor Mendel, proposed his laws of heredity and genes in 1866. It took several years later to realize the role of genes in evolution.

(ii) Every era is defined by its hero. Protein-based diagnosis and treatment strategies have had its share and have been exploited to the limits. It is time, we make the shift in the paradigm of our thought process. If not, our science will not make the next crucial leap necessary to improve the future medicine. I see miRNAs as the hero for the next generation-medicine. It has all the potential to serve as a good diagnostic marker. An ideal biomarker must be non-invasive, specific, sensitive, cost efficient, quantifiable, robust, translatable, and predictive. Scientists and physicians use biomarkers to monitor, screen, and diagnose various diseases. The commonly used disease-related biomarkers are primarily proteins; and to a lesser extent, lipids and electrolyte components. The major limitations of these markers are low specificity, sensitivity, and false positive results. These limitations are mainly to do with the biological property of the biomarker and to some extent the choice of the bio-physics based approaches. In my experience, it is every important to choose the correct tool to monitor expression profile of a biomarker candidate. There have been several new additions to the list of diagnostic tools like the flow cytometer or the use of magnetic beads, but all of these are only minor modifications to the use of the age-old antibody-based approaches. Where the old-time antibody-based diagnostic tools to monitor protein levels won, the modern approaches like adapting Raman tweezers failed. When it comes to proteins or other metabolites as biomarkers, I strongly believe we have plateaued to make any further improvements. To make new advancements in the field of diagnosis, we need to try out different classes of biomolecules. These are some of the reasons on why miRNAs will make an interesting biomarker candidate:

- (miRNAs) are endogenous, evolutionarily conserved, naturally abundant, relatively stable, small non-coding RNA molecules that function as post-transcriptional gene regulation in most biological and pathological processes.
- miRNAs are strongly dependent on physiological and pathological stimuli and reflects the functional state of a cell.
- The expression of miRNAs is tissue specific and can be used to identify specific disease conditions like specific tumor types and possibly their origin.

(iii) miRNAs have already been targeted to develop new-age therapeutics to treat a variety of disease conditions like cancers and heart diseases. The attractive characteristics in miRNAs that qualify it as a good target to developing therapeutics are that they are small, conserved, and with a known sequence. The miRNA-based therapeutics can either be miRNA mimics or molecules targeted at the specific miRNAs (anti-miRs). The miRNA-based therapeutics have shown promise in both animal and several pre-clinical studies. The current focus has been on enhancing the *in vivo* stability of the RNA molecule and defining the delivery methodology for the better.

From my experience, I strongly believe that this field of study is still at its infancy. There needs to be some form of a database that authenticate, organize, and present a detailed list of all the miRNAs discovered. Without this in place the study of miRNA, though interesting, is falling short of being tapped to its fullest potential. I believe that funding an organization to manage a miRNA-database and research is crucial to next-generation medicine. Also, it must be a non-bias and an apolitical organization to serve this purpose. The current approach (no names) was started with very good intentions but is falling short of performing at a competent level: poor correspondence and not up-to-date.

It is sad, that this knowledge about RNA, the creator of life, is not getting passed on as effectively as the other unimportant materials. Perhaps, this could be due to the system we have adopted and accepted it as the way of life. The leaders of the machinery called science must be apolitical but unfortunately, that is not the case. For example, a person who has never guided a graduate student is appointed as the Director of the graduate program committee; A professor who hardly shows up to work finds fault with their students not putting in enough time; or in a lighter vein, a captain of a club cricket team being a non-performer. In the near future, correcting such flaws will be able to promote science to the next level. Who knows, with time and acquiring in-depth knowledge on the different families of small ncRNAs may end up painting the eleventh dimension of life.

6

A Moron's Footnote to Controlling the Human Brain

The universe came into the present state because of the interactions they have undergone and are still undergoing, viz., attraction, repulsion, annihilation, and decay. The particles responsible for all these events are quarks and leptons.

To a common man, all forces we know and experience like gravity, friction, magnetism, and the rest are due to an unexplained phenomenon and is therefore attributed to the hand of God. Whereas in reality, this is due to the net result of particular interactions that occur without any direct contact; that is like the sun attracting earth and other planets. Magnetism and gravity are the best examples to show how things interact without touching one another. Interactions due to force career particles affect matter. As an example, electrons and protons have electric charge and therefore they produce and absorb the electromagnetic force carriers and on the other hand, neutrinos have no charge and hence cannot absorb or produce photons.

The creation and annihilations of quarks are due to an astronomical event that occurs during the last stages of stellar evolution of a massive star's life. It is the crucible from where the creation of all matter particles happen. It is now established that such an event occurs in our Milky Way galaxy once in 30 years or three times in a century; which actually coincides with the estimated gap between generations. This in-turn concurs with the attitude of mankind differing from one generation to the next. It is only my vague guess that a sudden spurt of a large number of quarks as a result of the explosion of massive stars could well be the reason for the systematic change in the ideology with time or generation. This may not be all true and the concept is still at its infancy, needing more research into particle physics; specifically in the field of super symmetry. The idea is simply based on the fact that the cosmic dust in the

© Springer Nature Switzerland AG 2019
S. M. Akula, *Essays on Life, Science and Society*,
https://doi.org/10.1007/978-3-030-28775-7_6

universe containing sub-atomic particles of varying angular momenta and charge coalesce to form atoms that we and what we see around are made up of. One thing to be kept in mind is that not all molecules are responsible for the life sustaining panspermia; they vary and every time there is an incidence of an explosion of a dying star, the particles may not have the same proportions to mix with the already existing particles. The soup is the same, only the taste is different. We now have established beyond doubt the numbers of protons, electrons, and neutrons in the molecules as well as the numbers of fermions and bosons that make up the protons and neutrons. However, we are still uncertain about the number of 'ghost particles' and their role in the creation of atoms.

Greed, lust, passion, sin, and desire were inherited at the beginning of evolution and overtime the intensity changed largely due to the belief in the unknown which was partly because of ignorance and the rest due to development; specifically at the time of industrial revolution. Human attributes are beyond comprehension, and if it is true that man can never meet God face-to-face, it is also equally true that man can never be able to decipher the behaviors of the human mind. And this perhaps is because we have neglected to look at the part played by the 'ghost particles'. The number of these particles that may inadvertently go into the process of forming atoms might determine the behavioral pattern of mankind over the centuries. Hypothetically speaking, a graph prepared would display that greed has overtaken and is progressing at an alarming rate compared to the rest of the human behavioral traits. At this stage, one has no other choice but to wait for the next supernova to occur and henceforth record the human behavior. The combination of subatomic particles in the creation of atoms may remain unchanged but it is the 'ghost particles' that remain the true culprits. Ghost particles can associate without exhibiting any perceivable change in the atoms. They are not only responsible for human behavior but are also responsible for all diverse ecological and biodiversity instability that we perceive, apart from manmade causes on the planet. The participation of fermions in the creation of atoms are well defined. The number of protons and neutrons that make atoms are well established in the molecules of say oxygen, hydrogen, carbon and the rest. On the contrary, the number of bosons required to glue the fermions are not well defined because unlike fermions, several bosons can exist in the same state. This is another area requiring research by biologists along with the physicists to find out whether this can lead to an answer to establish human behavioral pattern.

Should we lay hands on these aspects of science in the near future, we can be assured of controlling human behavior. In a way I perceive it, that is probably one solution to eradicating terrorism, genocide, apart from other terrible unfortunate deeds in such a way as to restore peace among our species and the environment as a whole. This is what science has to achieve to prove beyond a doubt that the unknowns can be unveiled. To sum it all up, particle physics like quantum mechanics, quantum electro dynamics, quantum field theory and quantum gravity are the stepping stones through which man can at last prove to the uninitiated, ignorant or arrogant masses that there is no such thing as unknowns or undefined. It just takes time. The flood gates of particle physics is now open for the physicists and the biologists to adapt it meaningfully to make living a meaningful and worthy one.

7

An Insiders' View!

Judging, making the right choice is one of life's important problems. Whether it be on decisions pertaining to life or work; making the call is like tying the bell around a cat's neck. One may even learn to appreciate the Henkin Semantics but can never claim to be the best judge. It is just a never-ending learning process. Science has its share of instances where good judgment is needed to make correct choices. Basically, you try to limit your errors in making the choices and this is possible by better understanding the situation at hand.

His name was Mohammad, an interesting grad student from Saudi Arabia who was pursuing his PhD in cancer biology. I happened to be one of his graduate dissertation committee members. He spent quite some time in my lab to work on a few of his experiments. He had a knack of starting conversations on some really new but interesting topics. One day, he asked me if I had ever watched the Oscar winner, Slumdog Millionaire? I said, I did not watch it and perhaps, would never watch it. He insisted that I watch it at least once as it was a fantastic one and an Oscar winner. I explained to him that a movie being an Oscar winner is not a good enough reason to watch it.

Mohammad insisted I give my reason why I was not going to watch an Oscar winner. I explained to him my very simple reasoning. In my opinion, any form of art must be appreciated for its beauty irrespective of the creator. Judging an art is subjective and relative to one's faculty of senses. Say there are a hundred movies created in a year; of which ten to fifteen movies are enjoyed by the vast majority. It would be sacrilegious to name one movie chosen by a panel of a set number of judges as the best one for that year. Even in the panel of judges, there is going to be differences of opinion and lobbying taking place

© Springer Nature Switzerland AG 2019
S. M. Akula, *Essays on Life, Science and Society*,
https://doi.org/10.1007/978-3-030-28775-7_7

and we know that lobbying is euphemism for bribing. In which case, how can one movie be named the best over the rest. It is an insult to the other creators. This is true for any form of art; be it, dance, music, painting, or martial arts performances. All creators of any form of art must be appreciated without any disparity. There should be days or events declared to glorify work of creators.

In my opinion, medals and trophies are better reserved for sports where it is objective and there is a clear winner. Even in sports, it is hard to judge diving, gymnastics and the likes as they too are subjective. When the measure of a success is subjective, you have wiggle room for politics. Politics in any form is the gateway to crime. Believe it or not, science has its own share of politics. It is deeply rooted in publishing in the field of science. Let me give you a few examples to justify my point. It was my dream to get my research published as a young scientist. About 25 years ago, author affiliated details were never sent to the reviewers and there was no such thing as publication fees. There was nobility in publication. When I was a graduate student at South Dakota State University things had slightly taken a change wherein there was still no publication fees, but the reviewers knew the source of the material. Within a few years, things had taken a big change; most journals started to charge a fee that ranged anywhere from $1000 to $3500. You have to pay for your scientific property to be published. There are really only a few exceptions like Journal of General Virology that do not tax the scientists. The functioning of the modern-day journals is ridiculous; perhaps, influenced by modern day medicine (pure business). Every day, there is an announcement of a new journal in the horizon because it is a money-making machine. To add to this, the authors got to provide a bunch of potential reviewers to review their own manuscript; which is a big hoax. There are so many instances where people just buy their way around. If this is the grounds for reality in publishing, then I shall let the reader's mind judge the manner on how grants are awarded. Such a democratic system has been groomed over time to a demonizing extent that there is no looking back. The only cool thing is those involved in this racket really believe in the fairness of this system. This perhaps is the major detriment to progress in science. Nepotism and cronyism drive scientific funding at all levels. The height of it all is inviting study section grant reviewers to present seminars in your department using departmental funds. This is a common practice followed in the name of networking. Isn't this a conflict of interest? There is no end to it and as the wise say, thieving will come to an end only when the thief realizes the act as being wrong.

For my part, I am a reluctant judge. This is perhaps due to the conformity instilled in me at a very early age. I would only serve as a judge if I had a good grasp on the subject matter. Even if I had an iota of doubt, I would decline the offer. Also, there is always a question in the back of my mind: "who are you to judge my work?" This kept a perineal check on my attitude towards judging other works. I tend to be gentle with my words, give an apt reason for the score, and highlight the best in each work. There will be no blanket statements or statements that demean a body of work. Demeaning other's work is a sign of authority and that is not a hallmark of a good scientist.

There are multitudes of questions and all arising from judging in science. But why is that so? Judging one's grant application decides the fate of a scientist. Judging at any level must be unbiased. Scientists who seldom spend time in the lab can become a great modern-day scientist; while, a good passionate scientist can be made to seem as a poor one; and all this by way of grants. How do we better shape the scientific community? How do we lead science into the next century? How do we develop the art of innovation? What is the solution to better drive this funding machinery? If the goals of the funding mechanism are wrong, all is wrong in science. This is the crux of the situation concerning any form of development in science. These are a few alternate ways of funding a scientist at work:

(i) Reduce the cap for total grant money: The total cap for grant money must never exceed $100,000 for 5 years. The lesser the money, the lesser the fraudulent practices. A meaningful good project can usually be completed with ease under $20,000. A smaller grant should not exceed $40,000 for 2 years.

(ii) Value the productivity of a scientist: The grant application must be backed up by a published manuscript(s) and not by preliminary data. This must be used not as a guideline but as a rule to one and all. Young scientists with a good track record of publication must be encouraged more than those without any.

(iii) No need for a recommendation letter: This idea of reference/recommendation letter should be banned. The argument against this one from the pundits would be as, "how can we judge a good investigator?" A scientist's publication record should speak for itself without a reference letter from a senior investigator; and this is a criminal approach. A reference letter is a moot point serving against an apolitical young and passionate mind. If you can't judge a chef by the meal prepared, you are just a poor judge or an imposter.

(iv) Reduce the cap on the total number of grants/scientist: At any given time, a scientist should never have more than 3 grants in a 5 year time period. The clock starts ticking from the time one receives their first grant. The only exception to this rule is when one's research leads to a patent that needs additional funds to develop the product.

(v) Make submission of a grant application easier: A scientist who can write a manuscript can definitely write a grant. A grant must not be more than a page and must be backed by the related manuscripts published from their laboratories on the second page. The one page write-up should describe their specific aims in brief, mention how it will benefit science, and list out possible manuscripts that the study will be able to generate. Keep it simple, after all, you are not dealing with Michael Faraday; all are equally intelligent and are a product of the pedagogical system of learning. Keeping it simple will avoid crass comments such as 'the person lacks grantsmanship'. This phrase is a highly discouraging remark and can only be used by a phony. There is nothing called grantsmanship. This is a word used liberally by the so-called judges when they do not have good logic to decline an applicant of their idea. Let me explain why? You either have it or not. If one has it, logically speaking, they should keep getting grants every time they applied for one. But that is not the case and I will let you mull on this.

(vi) Automate the grant scoring approach: We must eliminate gathering under the auspices of a dumb middleman and scoring grants around a table. It must be automated with lobbying considered a culpable offence. By reducing the cap for grant money and the number of grants, I envisage a significant drop in malpractice. The grant review process must be kept blind. This may take some time after all, they have been enjoying the benefits of this system for quite some time.

(vii) Do not decline the opportunity and still expect the scientist to perform well: It is foolhardy to expect a scientist to perform at the highest level without funding them. Never turn down scientists with passion to work and are actively publishing. This should be the motto of a funding agency.

(viii) Post-doctoral fellows must never be funded fully by the grants: The money from the grant must be used to perform the experiments proposed by the scientist and by the scientist themselves. In the process, they must train students for the next generation instead of wasting time by networking. A post-doctoral fellow must be a full-time employee hired by the university, hospital, or a company with all benefits as a faculty. They should be treated with respect and not be considered a slave worker. Based on the number of grants, the organization must hire these post-doctoral fellows by cost-sharing with the scientist. The goal of the organization is to educate students for the future and not syphon millions of dollars and show a profit.

(ix) What is a meaningful publication? To me, any publication is a good one. I have published in the top tier one's as well as in the low-profile journals. Going back to Mohammad, we have had several arguments over what a good work is. I consider every work published is worth it if you can only imagine from their shoes. When I was in India, getting my first publication in 'The Journal of Animal Sciences' was a great achievement as only I know what efforts went into it. It is foolish to ridicule that work. In the present times, those journals don't cross my mind but that does not mean the work carried by those journals are substandard. Instead of wasting time discussing good vs bad journals, we must look into creating a good set of journals with high standards. The current day journals are pure business modules. The goal is to mint money. I have paid up to $3500 to get my work published in some of the top peer-reviewed journals. But why collect this much money when they are already making much more from advertisements? The journals must do the following from their part to keep science at an affordable cost and more importantly provide a healthy competition:

- Keep the price affordable: Perhaps levying a fee of $200/article will keep it reasonable. This will avoid emergence of one or two new journals every week.
- Keep the review process blind.
- Give equal chance to rest of the world.

(x) Trust the work of each and every scientist: Ego is the major hurdle and to surpass it is another one. I always consider the best way to lead life is to compete against one's self and not against anyone else. If you mind your own business all is well.

These are a few ways to promote better science for the future. An aspiring young scientist should be taught how to do research and present their findings in terms of a manuscript and not to indulge in politics surrounding grants. Obtaining grants should be made simple: it should be a measure of publication and not whom you know in the committee. When we achieve it, we have science on the right path. Until then......we will be running second to Oscar movies!

8

Down the Line

Siva sat blissfully waiting for his wife to finish her second international phone call for the day. This was their Saturday routine; they would first call his parents and spend close to 45 min talking to them and then his wife would call her parents and have a long chat that would last the same length, while Siva watched TV. Today though, it was taking her a bit longer and Siva was getting restless. They were going to meet their friends at a restaurant for lunch and it was getting late. Before he could call out for her, Tamizh walked out of their bedroom and he noticed that though she was smiling, her eyes were downcast. He did not want to start a conversation just then and wanted to be on his way out. It was a silent drive to the restaurant.

Tamizh remembered the day she had first met Siva. She was in a car with her entire family driving from a different city to meet Siva. This was their formal introduction. If they liked each other then they would be married soon. The first time they met there were no sparks, but things felt normal and good. They both agreed to marry each other and conveyed their intent to their parents. Tamizh left her home, her country, to move to a foreign land that she had no interest to live in with Siva, who was also a stranger to her. She always thanked her stars that he turned out to be ok, unlike the horror stories she had heard of with fraudulent weddings. She smiled, she only called him ok. He was great a lot of the time, good sometimes, and not-so-good sometimes too, so ok was the best description she could come up with. In general, they were happy for the most part.

In the silent drive, memories rushed past Siva too. The first time he met Tamizh, that was 7 years go. It felt like another life time. In the airport when he walked away with her, he had sternly whispered to her, "no tears, you are

© Springer Nature Switzerland AG 2019
S. M. Akula, *Essays on Life, Science and Society*,
https://doi.org/10.1007/978-3-030-28775-7_8

leaving to be with your husband and for us to start a new life, it cannot start with tears," and he had seen Tamizh bite back her tears with downcast eyes. It was the same eyes he had seen in her today. For the most part he tried to keep her happy. He had a temper, he knew it, but he felt that he always tried his best to keep her interests in front of his. After all, it was she who had made the ultimate sacrifice of leaving her family behind to live with him, who was nothing more than a stranger when they had married.

Seven years! Tamizh thought to herself and more like 6 years of awkward conversations with friends and family. Every phone call started or ended with "any news?" Her parents were very direct with her and her in-laws were less direct, they paused, hemmed and hawed waiting to see if she would say anything. They thought they were smart but she knew. In the recent years, all the friends they had made were starting to pop out babies like it was the season. Some already had two kids and in any party or gathering she was prodded or teased. If she was ever caught talking to Siva alone, everyone Ahem'd as if they were going to make a baby right there. It was frustrating. What would Siva know, she thought.

Siva guessed what might have happened. It was not like he did not notice all those giggles and snide comments people made when they were together. He had seen enough Indian movies to know what families expect when two people got married, babies. That was the purpose of getting married, not to meet someone, fall in love, be happy, but to have babies. If they did not have a baby within the first 2 years then almost always they assumed that something was wrong with the girl. The boy never had to take responsibility. Thinking of this made his blood boil. He was a doctor, he knew how bodies worked. Their families were both educated and yet they succumbed to social pressures. He felt so helpless. He wanted to step in to help her but he knew that it would open a can of worms. He needed something solid to shut them up. Everyone up. He needed to come up with a good plan.

Tamizh, tensed up as she saw the restaurant sign come up. She was going to meet friends. Friends who would ask her why they were kidless after 7 years. She enjoyed their company very much until they asked those golden words, "anything brewing in your tummy?" She just hoped that they would not say anything to spoil the already horrible start. She could not even imagine that her own mother had suggested that she should go and talk to a doctor about having a child. Tamizh's face crinkled, as if they had not tried that 2 years ago. The first 5 years of their marriage life was so hectic that they had consciously made an effort to not have a child. They had to finish school, pay off debts, start a saving and start living the American dream before wanting to bring in a new being into the world. They did not want to raise a child in poverty. They were smart, they had had a plan. Once they started trying to have a child, nothing happened. Nothing. Every month her bloody guest arrived on time making her depressed sometimes. Doctor's had given them a green light and yet nothing happened. It made things worse when people had to address it so directly with her.

Siva, hoped that the Raju's whom they were meeting would not talk about it, if they did then he would have to do something drastic. Today, he would put an end to it. He knew the trouble his wife had been through. He had told his parents not to talk to her about it. He had to tell his in-laws the same too. It was not fair that people only addressed baby making questions to the wife only. It is not as if she was doing it all on her own and failing solo. He parked the car in front of the restaurant and told Tamizh to go in and he had a work call to make. Tamizh walked in to meet the Raju's.

Just as Tamizh started to cringe because her friend Malathy had asked her the question; Siva walked in. He sat down and looked at his friend Krishna and said, "I just got off the call from the doctor, I am impotent. So, stop asking Tamizh about children going forward." Tamizh's jaw dropped, she did not know what just happened, she looked at Siva in bewilderment and he just winked at her and whispered, "I called your dad and also told him before coming in." Tamizh laughed out loudly to the shock and surprise of Krishna and Malathy. The two of them laughed so hard and tried to get through their awkward lunch experience.

Gossip spread through their friend's circle so fast that they did not have to repeat it to anyone, soon no one troubled her with questions. They let nature take it's own course and miracles do happen and they were parents to a most gorgeous girl in 3 years. From the day at the restaurant Siva's and Tamizh's relationship bloomed to a different level. In Tamizh's head, her husband was not just ok but super all the time. He understood her through spoken and unspoken words and for him she was a super star for having tried to work through all this without even once mentioning or complaining. Having a child or becoming a parent did not define their marriage because it transcended beyond that.

9

What Is Intelligence?

Is it a measure of marks obtained by a student in a classroom? Or is it the measure of securing admissions in a privileged university? Or is it a measure of the number of degrees obtained? Or is it a measure of an IQ test score? Should this intelligence measure be used to define a person? When I was a young kid, I would have chosen the former reason, mostly due to ignorance. I was just an average kid in the class who dedicated more time to playing than studying. There were many intelligent kids in my class that I looked up to when it came to meeting educational or curriculum-based challenges. I spent lots of time in the playfield playing sports like tennis and cricket and my grades could never be compared to those of the intelligent kids in my class. When I was in my 10th grade, a change came upon me. Self-motivated, I decided to make some time and put some effort into studying everyday, but I promised myself that it would not be at the cost of my play time. That meant, I had to be awake for longer than usual hours. I did manage to do it and my grades improved considerably. In which case, I questioned, what is intelligence? Obtaining a good classroom score does not seem to be a sign of intelligence. The time spent on studying afterschool is directly proportional to the grades; and this result is statistically significant.

Along the same lines, securing admission into an Ivy league or other elite schools cannot be a sign of intelligence either. You may increase your chances of securing such seats by putting in time preparing for it or attend coaching classes. As we all know, in some countries you may purchase a medical or engineering admission by paying a capitation fee/donation. In some other countries, interviews decide the fate of a medical student applicant; and it is another form of corruption as you are going to let students with low scoring

S. M. Akula, *Essays on Life, Science and Society*,
https://doi.org/10.1007/978-3-030-28775-7_9

GPA's 3.3 and sometimes even lower getting admitted against GPA 4.0 scoring students. In all these cases, the history will never be revealed unless they run for the highest office. The prime example is President Bush, who is a Harvard alumnus and I will let you imagine the greatness of such schools.

I have also seen society revere people with multiple degrees. I question this, why? Is it because, a common man cannot obtain that many degrees? For my part, I respect such individuals for their interest to securing so many degrees but would not attribute to one's intelligence; it is more out of one's interest to prove a point. And it is definitely not a measure of IQ. It is estimated that a normal or an average intelligent person has an IQ of over 100. The percent of people that have obtained an IQ of over 100 is 49.93%; and most of this belong to the educated mass that have gone through the pedagogical manner of education. Once again, I strongly believe that this is a measure of success of a schooling system and not intelligence. After all, Einstein, a poor school student is said to have one of the highest IQ.

In my mind, intelligence is abstract, relative, and has no meaning to it. All are intelligent enough to perform skill sets at different levels based on the training received. For example, engineers and doctors are considered intelligent in the society. This is because of the following reasons:

(i) In many countries, only students who secure highest marks are eligible to pursue engineering or medical sciences for their higher education.
(ii) The training is said to be strenuous (especially for those who do not have a passion for it).
(iii) The service provided by them is considered as crucial to life or living.

But you must remember, these professions are not reserved for only the high-grade scoring kids in schools as I have seen students with GPA as low as 3.2 and 3.5 being admitted to such education. Beware, you never know who is certifying the power plant or treating you for an ailment. I have always considered referring a select few in schools as the brightest or intelligent as a form of racism. I have had my share of experience in my schooling days. By claiming a few students as intelligent over the rest only achieves two things: provides false confidence to those select few and inferiority complex to the rest.

In my opinion all are equally intelligent. Only that a few are book smart while many in the elementary schooling years lack that ability. This has to do with how they are raised at home and has nothing to do about intelligence. Here is my rationale and this has to do with a real life experience I would like to share. My brother and I completed our schooling in a Catholic school. My brother was an established sportsman and by his own admission, he was not good in studies. Even now we laugh about it as he is actively involved in educating young minds in the US. He was a fighter in the field and that ability did not simmer a bit even when outside. Once my dad was asked to meet with Rev. Brother Paul regarding his rendezvous in the school. My dad was used to this. As usual, he had a tough talk with my brother and visited Bro. Paul. He was the principal of our school and taught us moral science. A gentleman to the core who always had a serious face; smile met its defeat with him. Bro. Paul discussed the issue regarding my brother with my dad. I'm sure it was not easy for my dad; but I guess, by now he was well trained to handle the situation. My dad was about to leave his office, and this was exchanged between them as recollected by my dad:

Bro. Paul: Hey Mat! Sorry about that, but I had to share this with you.
My dad: I can understand (a bit let down)!
Bro. Paul: But I can say something. Your son will be successful in life.
My dad: What! After all this, but how? What makes you think that way, Bro. Paul?
Bro. Paul: Only tough minds shape this world. The softer ones accept life as it is and fade away.
My dad: If you say so and it comes true, I will be a happy man (my father felt emotionally relieved after the 30 min sermon). My dad narrated this long time ago and it is still engrailed in me. Perhaps, at that time Bro. Paul used this trick to alleviate my dad's spirit a bit as they were pals outside work. All said and done, it made sense to me. A great message from a fine soul.

There are three types of students formed during the elementary schooling years that rarely gets changed thereafter. Parents have a crucial role to play in the development of a child. Those three types are basically drafted by the parents in their living rooms. After all, the children are given no choice to select. These types are as follows:

- Type 1: The students are left carefree to fend for themselves in their homes. Basically, there is limited hands-on guidance provided to the young minds. The children develop for the good or bad by their own experiences.
- Type 2: The students are micro-managed to the extent that they are taught little bit of everything but not given enough time to master any particular stuff. Little bit of everything may include a sport, and music or dance. For these extracurricular activities, they are put in a coaching center. However,

the focus is primarily on studies. The parents strongly believe that their child must become an engineer or a doctor. There are no ifs and buts in this situation. The child's future is determined way before they were even conceived.

- Type 3: The students actively take part in sports; and they develop a passion for it. They may or may not do good in studies. In this case, there is no guarantee that the child is going to be a world beater in that specific sport.

Between the three types of students listed above, society always considers type 2 group of students to be as the brightest or intelligent. But I differ, and this is based on my experiences with life and reality. The greatest leaders, artists, scientists, and sportsmen all came from the type 1 and 3 group of students. There is an aggressive and active shaping of young minds happening during the elementary schooling ages between five till twelve. Evolution works at a micro level to select a few over the rest. You need to know what is losing to be a winner. You need to think independently to be a leader. You need to be fighter to take chances. True to being book smart, the chances of the group of students belonging to type 2 to becoming engineers, doctors, and so on is greater compared to the rest; however, seldom do they make the next leap to becoming a leader.

My take is simple: if one is intelligent than the others, then why go and attend a pedagogical way of schooling? You must be a Faraday, or Nabokov! And this is one of the reasons why I hate these award mechanisms kept in schools. In fact, I would go one step further to remove all these dubious awards being presented to the performing students. By this way, you refrain from infusing a kind of superior feeling to a select few at a very young age. An attitude of that kind is not the right precedence to set up a child's future because such individuals tend to become vulnerable in life towards handling losses. On the same lines, eliminating of such award ceremonies may in fact limit pervading inferiority complex in the rest. If you are good, why do you need an award; especially in the field of education which should be accessible to everyone at the same level. The goal is to create a prosperous generation for the future and not a few select 'so-called' intelligent one's who may never fit the society. What good is to have an intellect that is a mis-fit in the society and that is the plain fact why these clever students from the schools seldom make it to the top in a society. The entire schooling system must be an exciting fun filled experience where young minds with identical interests are identified, let to work in groups to think and perhaps develop novel and simple solutions for complex problems. The key for a good education system is not to identify and segregate book smart children into small groups; rather, identify and nurture each and every possible talent and It should all start from the very early days in a school. Talent does not mean securing good marks in science and mathematics alone. Parents must realize it and pass it on effectively to the next generation. Doctors and engineers alone cannot make life worth living. Some may claim that people belonging to these two professions play a crucial part in life and that is foolhardy. An engineer depends on mathematicians, physicists, geologists, environmentalists and others to design a structure and so does a doctor who will cease to prescribe a medicine if not for the life scientists belonging to varied fields like bacteriology, biochemistry, chemistry, genetics, microbiology, molecular biology, pharmacology, physiology, and so forth to develop it. They are just a part of life. Society has to realize this and accordingly define the economics around each and every profession. No one is superior than the other. Every profession is dependent on the other to make life frolic. Talent can range from various forms of arts (different forms of writing, painting, singing, playing instruments, acting, dancing, speaking), sports, cooking, carpentry, politics, business, science, cleaning, organizing, etc. Otherwise, you will have students pursue higher education due to peer pressure and not a passion for the field which would probably render them being less rapt or qualified in the profession. As Einstein once mentioned that it would be a futile attempt to measure the smartness of a fish by its ability to climb a tree. For my part, all are intelligent and the measure of success in life is how happy one is and how successful one is in contributing to the society independent of their profession.

Printed in the United States
By Bookmasters